Bulletin 376
Development Policy & Practice

Access of the poor to agricultural services

The role of farmers' organizations in social inclusion

Bertus Wennink, Suzanne Nederlof and Willem Heemskerk (eds.)

Contents

5	Acknowledgements
7	Foreword
9	Abbreviations
11	Introduction

Part I Enhancing agricultural service provision for the rural poor

17	1 Background
25	2 Farmers' organizations in Sub-Saharan Africa
35	3 Analytical framework
39	4 Methodology of case studies
41	5 Farmers' organizations and social inclusive service provision
61	6 Concluding remarks: towards a strategy for social inclusion
65	References

Part II Case studies on the role of farmers' organizations in accessing services

71	I INGABO's role in pro-poor service provision in Rwanda *Jean Damascène Nyamwasa and Bertus Wennink*
87	II MVIWATA's role in pro-poor service provision in Tanzania *Stephen Ruvuga, Richard Masandika and Willem Heemskerk*
105	III UCPC's role in pro-poor service provision in Benin *Clarisse Tama-Imorou, Bertus Wennink, and E. Suzanne Nederlof*
125	IV KILICAFE's role in pro-poor service provision in Tanzania *Adolph Kumburu and Willem Heemskerk*
139	V ACooBéPA's role in pro-poor service provision in Benin *Clarisse Tama-Imorou1 and Bertus Wennink*
155	About the authors

Acknowledgements

This bulletin is a joint effort of farmers' organizations, Non-Governmental Organizations (NGO) and KIT researchers. Comparing experiences in the field of social inclusion in farmers' organizations has greatly enriched insights towards enhanced access to services.

We thank the staff and members of the following organizations for their inspiring input, contributions and feedback:
- The *Association des Coopératives Béninoises de Planteurs d'Anacardier* (ACooBéPA) in the Ouèssè and Tchaourou districts in Benin;
- The Association of Kilimanjaro Specialty Coffee Growers (KILICAFE) in Tanzania;
- The National Network of Farmer Groups (MVIWATA) and MVIWAMO, an intermediate-level network in Monduli District, in Tanzania;
- The *Syndicat Rwandais des Agriculteurs et des Eleveurs* (INGABO) in Rwanda and the *Réseau des Organisations Paysannes du Rwanda* (ROPARWA), and
- The *Unions Communales des Producteurs de Coton* (UCPCs) of Kandi, Boukoumbé and Djidja districts Benin and the *Fédération des Unions des Producteurs du Bénin* (FUPRO) in Benin;

Special thanks go to the co-authors of the case studies who accompagnied us all the way. It is thanks to them that this bulletin is rich in information and provides an insight into the life of farmers' organizations in Benin, Rwanda and Tanzania.

We would also like to thank the Directorate-General for International Cooperation (DGIS) of the Dutch Ministry of Foreign Affairs, who made the field studies as well as the editing and publishing of this bulletin possible.

The editors would also like to thank Jacob Kampen for a critical review of the draft versions of this bulletin, Barbara Shapland for the language editing, and Sarah Simpson for the final touch.

Bertus Wennink, Suzanne Nederlof and Willem Heemskerk

Foreword

Noah Mofuna is a farmer from Central-Togo. Some years back he, together with ten other cowpea producers from the same village, joined in a Farmers' Organization for Cowpea Production (FOCP). That year the group sold their cowpea for a good price through the intermediary of an NGO. FOCP also managed to convince the local authorities to help them with the transport of the produce and they contacted another agricultural service provider to assist them in testing improved cowpea varieties. Noah is very proud to belong to the group. People in the village often turn to him for advice. Yet, it is not possible for everybody to join. For example, Noah's neighbour, Sarah Lawat, a 60-year old widow without children whose husband died from HIV/AIDS, cannot pay the contribution that group members collect (even though Noah considers it a small amount, i.e., the equivalent of 50 euro cents a week). She also does not have the strength to help them with the work in the common group field; she barely manages to cultivate her own small village plots. Noah reflects: 'farmer groups are good for me and those who are like me, but what about Sarah, and the other less-endowed fellow villagers? How can they lobby for better prices, contact NGOs and get the local authorities to improve infrastructure?'[1]

Sarah is not alone, there are many others who, like her, do not join farmers' organizations, sometimes because they are weak, female, old, or poor and in other cases because they are nomads, herders, migrants, from a minority ethnic group or ill (HIV/ AIDS, malaria etc. are all too common amongst rural poor). In yet other situations, sometimes farmers who do not grow state-supported market-oriented commodities are excluded from access to agricultural services such as extension and/or input supply. As a result the specific needs of these categories of farmers often are not provided for or defended. It is these issues that this bulletin is about: under what conditions might people like Sarah also benefit from agricultural services and what could be the role of farmers' organizations in this endeavour? Is it even possible that farmer organizations facilitate access to services for the poorest? What is needed to make deprived farmers benefit more, and what strategies would enhance social inclusion?

More specifically, this bulletin attempts to address farmer organization-related issues with the ultimate goal of developing guidelines for a pro-active strategy for social inclusion of disadvantaged groups or individuals in farmers' organizations to enhance their improved access to agricultural services. Development practitioners and other players in the field of farmer empowerment and farmer organizations as well as policy makers who could use these guidelines are the intended audience of this publication.

Note

1   This is a fictional case.

Abbreviations

ACooBéPA	*Association des Coopératives Béninoises de Planteurs d'Anacardier*
ADIAB	*Association des Distributeurs d'Intrants Agricoles du Bénin*
AGM	Annual General Meeting
AGROP	*Association des Groupements de Producteurs*
AIC	*Association Interprofessionnelle du Coton*
AKSCG	Association of Kilimanjaro Specialty Coffee Growers
AMSDP	Agricultural Marketing Sector Development Programme
ANPC	*Association Nationale des Producteurs de Coton*
ASDP	Agriculture Sector Development Programme
CAGIA	*Coopérative d'Achat et de Gestion des Intrants Agricoles*
CARDER	*Centre d'Action Régional pour le Développement Rural*
CBO	Community Based Organization
CeCPA	*Centre Communal pour la Promotion Agricole*
CeRPA	*Centre Régional pour la Promotion Agricole*
CLCAM	*Caisse Locale de Crédit Agricole et Mutuel*
CLECAM	*Caisse Locale d'Epargne et de Crédits Agricoles Mutuels*
CNIA	*Centre National pour l'Insemination Artificielle*
CPU	Central Pulping Unit
CRA-CF	*Centre de Recherche Agricole – Coton et Fibres*
CSPR	*Central de Sécurisation de Payement et de Recouvrement*
DEDRAS	*Organisation pour le Développement Durable, le Renforcement et l'Auto-promotion des Structures communautaires*
DFF	District Farmer Fora
EAFF	East African Farmers' Federation
EZCORE	Eastern Zone Client-Oriented Research and Extension
FANAPRA	*Fédération Nationale des Producteurs Agricoles*
FBG	Farmer Business Group
FBO	Farmer Based Organization
FERWATHE	*Fédération Rwandaise des Producteurs du Thé*
FF	Farmer Fora
FFS	Farmer Field School
FG	Farmer Group

FOCP	Farmers' Organizations for Cowpea Production
FUPRO	*Fédération des Unions des Producteurs du Bénin*
GDP	Gross Domestic Product
GPA	*Groupements des Planteurs d'Anacardier*
GV	*Groupement Villageois*
GVPC	*Groupement Villageois des Producteurs du Coton*
IFAD	International Fund for Agricultural Development
IFAP	International Federation of Agricultural Producers
IMBARAGA	*Syndicat des Agriculteurs et Eleveurs du Rwanda*
INADES	*Institut Africain pour le Développement Economique et Social*
INGABO	*Syndicat Rwandais des Agriculteurs et des Eleveurs*
INRAB	*Institut National des Recherches Agricoles du Bénin*
ISAR	*Institut des Sciences Agronomiques du Rwanda*
KILICAFE	Association of Kilimanjaro Specialty Coffee Growers in Tanzania
KIT	Royal Tropical Institute
KNCU	Kilimanjaro Native Cooperative Union
MoU	Memorandum of Understanding
MT	Metric Tons
MVIWAMO	*Mtandao wa Vikundi vya Wakulima wa Wilaya ya Monduli*
MVIWATA	*Mtandao wa Vikundi vya Wakulima ya Tanzania*
NGO	Non-Governmental Organization
NRMC	Natural Resource Management Committee
PADEP	Participatory Agricultural Development Programme
PADSE	*Projet d'Amélioration et de Diversification des Systèmes d'Exploitation*
PELUM	Participatory Ecological Land Use Management
ROPARWA	*Réseau des Organisations Paysannes du Rwanda*
SC	Steering Committee
SCAA	Specialty Coffee Association of America
SIP	*Société Indigène de Prévoyance*
SMDR	*Société Mutuelle de Développement Rural*
SONAPRA	*Société Nationale pour la Promotion Agricole*
SWOT	Strengths Weaknesses Opportunities and Threats
TACRI	Tanzania Coffee Research Institute
TCB	Tanzania Coffee Board
TCGA	Tanganyika Coffee Growers' Association
TNS	Technoserve
ToT	Training of Trainers
TWT	Taylor Winch Tanzania Ltd
UCP	*Union Communale des Producteurs*
UCPC	*Union Communale des Producteurs de Cotton*
UDP	*Union Départementale des Producteurs*
UDPC	*Union Départementale des Producteurs de Coton*
UEEB	*Union des Eglises Evangéliques du Bénin*
ULGPA	*Union Locale des Groupements des Planteurs d'Anacardier*
USPP	*Union Sous-Préfectorale des Producteurs*
WFF	Ward Farmer Fora

Introduction

Justification

Farmers' organizations today play a much more prominent role in agricultural policy formulation and implementation in Sub-Saharan Africa than ever before. In a context of liberalization of the agricultural sector, privatization of delivery of goods and services, and political democratization, farmers' organizations claim their stake and are recognized as key stakeholders in rural development. For both the public and private sector, effective farmers' organizations present important opportunities such as: providing research and extension services to farmers and organizing the purchase of inputs and sale of products on a more cost-effective basis; mobilizing resources for local development; and representing the interests and collective voice of farmers in development fora (Bosc *et al.*, 2003; Chirwa *et al.*, 2005).

Farmers' organizations distinguish themselves from other public and private sector organizations through their membership base. These are rural organizations whose members share a common interest. Farmers' organizations are basically democratic organizations, often with a strong 'grass roots' basis that (on behalf of their members) may apply different approaches in their relations and interactions with other stakeholders in the agricultural sector. These approaches are based on a combination of style (cooperative or confrontational) and basis (evidence and science-based or interest and value-based). The resulting respective functions: advisory and lobbying (cooperational) and advocacy and activism (confrontational), are in the interests of an organization's members. This results in a collective voice of the members through representation, and improved services through (reorientation and/or provision of technical and economical services) that more effectively respond to members' needs (Bosc *et al.*, 2003).

The services that are being provided to members, whether by farmers' organizations themselves or by third parties, include knowledge services such as agricultural research, advisory (extension and technology dissemination) and other types of farmer training. Such services are increasingly considered key factors for advancing rural development. However, improving agricultural practices and processes through effective knowledge application, requires two basic pre-conditions: access of farmers to appropriate knowledge sources and services, and a conducive context that incites knowledge application (World Bank, 2006). Farmers' organizations can play a key role in agricultural innovation, since they have the capacity to pool, aggregate and disseminate knowledge and information (Collion and Rondot, 1998). Moreover, they are increasingly positioned in both service networks and supply chains to coordinate activities and promote an enabling environment for innovation.

Poverty in Sub-Saharan Africa is still mostly a rural phenomenon despite rapid urbanization; more than 70% of the poor live in rural areas (IFAD, 2007). Agriculture remains a key sector for alleviating poverty in rural areas and has received renewed attention on the development cooperation agenda (see for example DFID, 2005; OECD, 2006; and World Bank, 2007). Agriculture is still the main economic activity for most rural people; it remains an important source of income for farmer households and contributes to sustainable financing of social-sector services (Irz *et al.*, 2001). Poverty is the result of economic, social and political processes that often reinforce each other. Meagre assets, difficulties in grasping the opportunities that are potentially available and exercising countervailing power, often related to the policy and social context, are determining factors in the situation of the rural poor. Vulnerability to events that are out of their control often exacerbates their poverty situation (World Bank, 2001).

Strategies aimed at alleviating poverty therefore include three key elements: identifying opportunities (e.g., access to natural resources, markets and service provision to build up assets); facilitating empowerment (e.g., participation by the poor in political processes and decision-making); and, enhancing security (Ibid). Social *inclusion* of service provision essentially refers to the access to services by the most vulnerable farmers in rural society. Access to knowledge is required for growth, but if the context is not right, or if farmers' access is not inclusive (of the rural poor), such growth will not lead to well-balanced development and certainly not to pro-poor development. Social *exclusion* leads to research and development agendas which do not include the priorities of the poor, resulting in constrained access by the poor to appropriate knowledge and hence to their exclusion from economic and social progress.

Farmers' organizations are increasingly involved in orienting services towards the specific needs of their members and/or providing these services themselves. However, although the role of farmers' organizations in Sub-Saharan Africa is rapidly increasing in importance, there are significant risks that individual farmers and/or groups are being excluded from these services. There are also many farmers who do not join farmers' organizations. Sometimes this is because they are particularly poor or belong to vulnerable groups, such as female-headed households and widows, and in other cases because they are from a minority social or ethnic group, or disabled (HIV/AIDS-affected households are all too common amongst the rural poor). In other situations, subsistence farmers who do not produce marketable commodities may have difficulties in becoming members of farmers' organizations and therefore in accessing relevant agricultural services. As a result, the specific needs of these categories of farmers are often not provided for, or defended, and they are excluded from effective service provision. Social exclusion or inclusion in service provision unfolds through the interactions between the different stakeholders involved, including farmers' organizations and agricultural service providers in the public, private and 'third' sector[1], and is therefore strongly related to the institutional context.

Case studies drawn from experiences in Sub-Saharan Africa show that agricultural research and advisory services are increasingly channelled through farmers' organizations (Wennink and Heemskerk, 2006). Farmers' organizations that provide these services themselves, are often directly supported by NGOs and donors, and are increasingly being contracted to provide advisory services by the public sector, and sometimes also by the private sector. In addition, farmer groups and organizations increasingly voice their members' concerns and have a say in issues that impact farmers' livelihoods. These same case studies also show the discrepancy in dealing with service provision between more inclusive, mostly smaller, community-based farmer groups (those oriented towards enhanced livelihoods), and often less inclusive and larger commodity-based producer organizations (supply-chain oriented). This is the main focus of this bulletin: the role of farmers' organizations in facilitating access by the poorest farmers to agricultural services, and under which conditions such organizations can enhance social inclusion.

This bulletin

This bulletin focuses on two major questions: How do the poorest of the poor gain access to, and benefit from, agricultural services? What is the role of farmers' organizations in socially inclusive access to these services, and to what extent is membership of the farmers' organizations a determining factor for this? Answers to these questions will hopefully allow guidelines and strategies to be defined for improving the livelihoods of the rural poor by enhancing their access to agricultural services, including through farmers' organizations.

The subject of inclusion of farmers, their groups and organizations in setting research agendas, extension priorities and in carrying out field experiments is not addressed in this bulletin (for further information, see Nederlof, 2006). However, the importance of farmers' organizations in facilitating socially inclusive access to agricultural services, and the relevance of including different categories of farmers in farmers' organizations, as well as their representative roles, forms the main topic of this bulletin. Who joins farmers' organizations, and why? What is the impact of group rules, procedures and mechanisms (of adherence, participation and relations with the surrounding environments) on the membership? What role do the members play within the organization? Which individuals and groups do the farmers' organizations represent in addition to their members? What is needed to help ensure that agricultural services are not exclusively aimed at the relatively richer farmers?

More specifically, this bulletin attempts to address these questions with the ultimate goals of developing guidelines for a proactive strategy for social inclusion of disadvantaged groups or individuals in enhanced access to agricultural services, through farmers' organizations. Development practitioners and other players in the field of farmer empowerment and farmers' organizations, as well as policymakers who could use these guidelines form the intended audience of this publication.

This bulletin is divided into two parts. Part I is an analysis of social inclusion and the role of farmers' organizations in access to agricultural services; Part II contains a description of the case studies (on which Part I is based). The first part discussed social inclusion within a context of poverty, sustainable livelihoods and empowerment. The context of farmers' organizations and their roles in obtaining access to service provision is described. The question of social inclusion of disadvantaged and vulnerable farmers within farmers' organizations is also addressed. A tentative conceptual framework consisting of issues relevant to an active social inclusion strategy is presented next. Experiences reported in the literature, as well as emerging ideas from several case studies (reported in Part II) that were developed simultaneously, were used when developing this framework. The conceptual framework should therefore be considered an outcome of the case studies as well as an input. The concluding remarks discuss the following issues: the policy context and enabling environment for pro-poor development; the nature (socially inclusive or exclusive) of farmers' organizations; the consequences of such social exclusiveness; the role of farmers' organizations in inclusion or exclusion in agricultural services; and the way in which farmers' organizations can enhance social inclusion in services.

Part II of this bulletin describes the case studies on farmers' organizations in Tanzania, Rwanda and Benin that were used for the analysis of social inclusion.

Note

1   The third sector comprises organizations that are not fully in the public or private sector, such as voluntary organizations and community groups.

Part I
Enhancing agricultural service provision for the rural poor

1 Background

Poverty and the poor

Poverty is the result of exclusion from economic, political and social processes, and for that reason, promoting opportunity (such as improving market functioning and stimulating economic growth) is important in fighting poverty. However that alone is not enough: poverty is also influenced by the unequal distribution of power and by social norms, values and customary practices (e.g., taboos on crop management practices, levelling mechanisms[1] and/or local/traditional justice), which might lead to exclusion. Therefore, empowerment of 'the poor' is also important in fighting poverty. A third pathway towards alleviating poverty consists of enhancing security by reducing risks of vulnerability, which can be both natural, man-made and/or economic (World Bank, 2001).

It therefore follows that there are several dimensions to 'being poor', such as:
1. lacking adequate food and shelter (due to no, or very low, income), poor access to education and health services, and other deprivations that keep a person from leading the kind of life that everyone values;
2. facing extreme vulnerability to ill health, economic dislocation and natural disasters; and,
3. being exposed to poor treatment by state institutions and society at large, and being powerless to influence key decisions affecting one's life.

Economic growth and income are on the rise in developing countries (DfID, 2004). Yet, in general, those who are already richer benefit relatively more than those who are poor. It is important to point out that economic growth does not automatically lead to overall development and poverty alleviation (Øyen, 2001)[2], but may sometimes even lead to greater poverty[3].

'Empowerment' and 'security' are not the only means to achieve 'economic growth' (Shirbekk and St.Clair, 2001). This bulletin adopts the multi-dimensional perspective of poverty, which development practitioners recognize in real life. To paraphrase Shirbekk and St.Clair (ibid: p. 15) who refer to Sen (1981):

> *Development ought not to be conceptualized as the achievement of modernization, industrialization and economic growth, but as the expansion of people's capabilities and functionings.*[4]

In order to consider the various dimensions of poverty and to put the poor in the centre, it is useful to adopt a sustainable livelihoods perspective[5] (for more information, see IDS, 2006). A sustainable livelihoods perspective focuses on:
1. a holistic understanding of access to, and control over, capital (natural, financial, social, human and physical);
2. the context of vulnerability for the poor; and,
3. processes, institutions and policies at all levels that help or constrain people to use their different kinds of capital for improved livelihoods (DfID/FAO, 2000).

Such a perspective helps us find ways to enhance a policy and institutional environment, to better support poor people's livelihoods while building on their strengths. Poor people have their own strategies to secure their livelihoods depending on such factors as their socioeconomic status, education and local knowledge, ethnicity and the stage in the life cycle of the household (Messer and Townsley, 2003).

Social exclusion and inclusion

It is important to have a clear understanding of what social inclusion means as it eventually determines how to develop useful strategies for enhancing social inclusion. In the case of disadvantaged and vulnerable farmers this means: to understand the way they access agricultural services, whether through actual membership of farmers' organizations or through indirect representation by farmers' organizations. When talking about social inclusion one cannot escape discussing social exclusion. A social exclusion perspective focuses on two sets of barriers to alleviate poverty, namely:
1. social relations (or lack thereof) that exclude people; and,
2. restricted access to institutions and organizations that matter for poverty alleviation, citizenship and rights (Beall and Piron, 2005).

Hence, social exclusion might be a reason why the poorest of the poor have less access to, and participate less often in, farmers' organizations, and thus have less access to agricultural services. The most common definition for social exclusion is probably the one used by Eames and Adebowale (2002: p. 3):

> *Social exclusion is the condition of communities, groups and individuals who are economically and/or socially disadvantaged.*

According to this definition, categories of socially excluded people include those living on (relatively) low(er) incomes and people from minority ethnic communities. However, a variety of different definitions for social exclusion are being used (Farrington, 2002).

Differences in the way in which social exclusion is defined relate to:
- Exclusion being considered as either a condition or the process itself.
- The people it affects. Exclusion affects individuals (through for example lack of social capital[6]), certain groups or everyone.
- The environment surrounding people. People's environment can constitute barriers and lead to exclusion from labour markets, breakdown of 'social systems', and/or (lack of) resources. Using the theory of social capital, social exclusion is considered an important cause of poverty (Toye and Infanti, 2004).

The following elements are essential for explaining the dimensions of social exclusion:
- Recognize the dynamic nature of social exclusion. Therefore, similar to the aforementioned understanding of poverty, exclusion can be considered a process and not a (fixed/static) condition. This helps to understand the causes of exclusion and consequently to develop a strategy for addressing these causes and including the 'poorest of the resource-poor farmers' in farmers' organizations to improve access to services and thereby their livelihoods. It also emphasizes the inter-connectivity of the causes of exclusion.
- In the same vein, social relationships are important in exclusion processes; this explains this bulletin's focus on social capital (see also Heemskerk and Wennink, 2004) and, as a result the need for active participation of both individuals and organized groups. After all, social exclusion affects each individual, as well as society as a whole.
- Along the same lines, exclusion not only involves the more material aspects of exclusion, but also the exclusion from social, economic, institutional, territorial and symbolic reference systems (for a discussion on these systems, see Farrington, 2002), and also includes economic, political or cultural aspects.

Shookner (2002) created a tool that he calls 'an inclusion lens'. This tool helps to understand inclusion (who are to be included, who will benefit, what are the measures that would promote inclusion) and to develop an action plan. The inclusion lens is also a tool for analyzing legislation, policies, programmes and practices, to determine whether they promote the social and economic inclusion of the poorer individuals, groups and communities. The 'inclusion lens' is a list of dimensions and elements that favour inclusion (see Table 1).

Table 1: An inclusion lens

Dimensions	Element of inclusion
Cultural	Valuing contributions of both women and men to society, recognizing differences, valuing diversity, positive identity, and anti-racist education.
Economic	Adequate income for basic needs and participation in society, poverty eradication, employment, capability for personal development, personal security, sustainable development, reducing disparities, giving value and support care.
Functional	Ability to participate, opportunities for personal development, valued social roles, and recognizing competence.
Participatory	Empowerment, freedom to choose, contribution to community, access to programmes, resources and capacity to support participation, involvement in decision-making, and social action.
Physical	Access to public places and community resources, physical proximity and opportunities for interaction, healthy/supportive environments, access to transportation, and sustainability.
Political	Affirmation of human rights, enabling policies and legislation, social protection for vulnerable groups, removing systemic barriers, willingness to take action, long-term view, multi-dimensional, citizen participation, and transparent decision-making.
Relational	Belonging, social proximity, respect, recognition, cooperation, solidarity, family support, and access to resources.
Structural	Entitlements, access to programmes, transparent pathways to access, affirmative action, community capacity building, inter-departmental links, inter-governmental links, accountability, open channels of communication, options for change, and flexibility.

Source: Shookner, 2002.

It is important to realize that social inclusion is not necessarily the solution to social exclusion (Beall and Piron, 2005) since some groups may deliberately choose to remain outside the 'mainstream'. In other words: some groups 'self-exclude' themselves (Toye and Infanti (2004: p. 17), paraphrasing Jackson (2001)):

> *An inclusive group (or society as a whole) is characterized by a widely shared social experience and active participation, by a broad equality of opportunities and life chances for individuals and by the achievement of a basic level of well-being for all members (/citizens).*

A strategy towards social inclusion includes an approach of handing over the necessary means to poor people or empowering poor people with knowledge or other resources to give them the opportunity to generate their own tools to achieve enhanced livelihoods.

Empowerment and voice

The issue of inclusiveness of farmers' organizations and service provision to their members, and who benefits from these services, is closely related to the people's level of empowerment. Empowerment is about people taking increased control over their lives and destiny. In this bulletin, empowerment refers to (Kabeer, 2001):

> *The expansion in people's ability to make strategic life choices in a context where this ability was previously denied to them.*

Barlett (2004) presents a simple model of the steps involved in a transformation process towards empowerment (see Figure 1). All three steps are needed: generally a change in means establishes the potential for a change in process, and a change in process allows a change in ends. In turn, a change in ends might in itself bring about a further change in means, etc.

Figure 1: A transformation model of empowerment

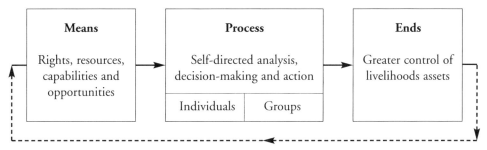

Source: Barlett, 2004.

Means can involve many things, ranging from national legal and political systems to the resources and the skills of people themselves. Training, establishing farmers' organizations and linking them to stakeholders, all contribute to changing the means for empowerment.

Once people have increased their means it is important that they decide what to do with them. Only when people analyze for themselves, make their own decisions and determine their actions, can one state that they are really empowered. In some cases individuals decide, but in other cases it is a group decision; therefore the answer to the question of 'who decides' is relevant to understanding who is empowered.

The end (i.e., achievement of empowerment) involves an increased influence of people over the conditions (quality, security, dwelling etc.) of their lives. So the question then is: which conditions are people trying to change when they become empowered? One possible answer concerns changes in relationships (e.g., women in relation to men, poor in relation to rich, or civil society in relation to the state). It is difficult for 'outsiders' such as development cooperation agencies to change these types of conditions. Increased control over livelihoods and assets is another possible response. People gain greater control over their human capital (e.g., knowledge, skills, health, etc.), social capital (within groups and networks), natural capital (e.g., land and water), physical capital (e.g., houses, roads and sanitation), and financial capital (e.g., savings, credit, wage rates). It is important to look at the livelihoods approach from the perspective of the people involved.

The question of what people want must be answered by themselves. They need to speak out and, in order to affect change, they also need to be heard and listened to, and their aspirations must be taken into account. In other words, people need to exercise their 'voice' (Bebbington and Thompson, 2004). Voice is therefore considered to be an important means of improving the responsiveness and inclusiveness of services. In a more pluralistic context of service provision, service users can leave and obtain their services from another provider, thus forcing service providers to improve, but in many other contexts this is not possible. Often there are either no alternative providers, users have little power and/or clients are in 'patronage networks (Goetz and Gaventa, 2001). In this bulletin, 'voice' is defined as (Ibid):

> *The range of measures used by civil society actors to put pressure on service providers to demand better service outcome.*

On the basis of case studies, Goetz and Gaventa (2001) distinguished three types of initiatives for making services more responsive (see Table 2):

However, the characteristics of both users and the services involved influence the way in which users, including the poorest, exercise their voice and the way in which providers respond. This concerns the nature of client relationships; the geographical dispersion or concentration of users; the social status of users; the costs of services; the market mechanisms involved; and, the way services are being delivered (Ibid). The most disadvantaged farmers, their organizations and agricultural service providers, therefore need to be involved to make knowledge work for more inclusive development. Exclusion of the poorest from the innovation system will probably prohibit rapid development of relevant knowledge and adoptable technologies, while specific knowledge of excluded groups will not be used. The agricultural innovation system perspective therefore refers to the need to involve all actors, including the most disadvantaged, in an innovation system that contributes to inclusive development (World Bank, 2006).

Table 2: Types of initiatives for making services more responsive

Citizens' initiatives	Joint civil society and public-sector initiatives	Public-sector initiatives
Awareness-raising and capacity building for mobilization	Implementation and precedent-setting (including partnerships)	Consultation on users' needs (for policies and services)
Information generation (research for advocacy)	Auditing	Setting standards
Lobbying to influence planning and policy formulation	Joint management of sector programmes	Incentives, sanctions and performance measures
Citizen-based monitoring and evaluation	Government frameworks for participatory planning	Service delivery 'ethos' in organizational culture Accessible (government) information and services New rights for citizens or clients

Source: Goetz and Gaventa (2001).

Notes

1. Levelling mechanisms aim to even out the distribution of wealth (Shrestha, 1990) and imply that people do not want to publicly display their wealth. Hence, people do not want to show too obviously if they are much richer than their neighbours.
2. Critics – for example Øyen (2001) – state that this is not clear in the World Development Report 2000/2001.
3. The USA, and more recently India, are probably the best examples of countries with high economic growth, but also extreme poverty.
4. Capabilities refer to what people can or cannot do, and functionings refer to what people actually do, or do not do (health, food, education).
5. Another, yet complementary perspective, is the 'Rights-based Approach' (see for example Mukhopadhyay and Sultan, 2005).
6. For more information about social capital, see Heemskerk and Wennink (2004). The term 'social capital' originates from the work of Bourdieu, who distinguishes between three forms of capital: economic capital, cultural capital and social capital. He defines social capital as 'the aggregate of the actual or potential resources which are linked to possession of a durable network of more or less institutionalized relationships of mutual acquaintances and recognition'. (http://en.wikipedia.org/wiki/Social_capital). Using the theory of social capital, social exclusion is considered an important cause of poverty (Toye and Infanti, 2004). The theory of social capital gained popularity after Robert Putnam wrote a book entitled 'Bowling Alone' (2000). He analyzed what he has called the collapse of social capital in the USA. He distinguishes between 'bonding' and 'bridging' capital. Bonding occurs when you are socializing with people who are like you: same age, same race, same religion etc. But in order to create peaceful societies in a diverse multi-ethnic country, one needs to

have a second kind of social capital: bridging. Bridging is what you do when you make friends with people who are not like you, e.g., supporters from another football team. Putman argues that those two kinds of social capital, bonding and bridging, reinforce each other mutually. (From: http://en.wikipedia.org/wiki/Robert_Putnam).

2 Farmers' organizations in Sub-Saharan Africa

Characteristics of farmers' organizations

Where institutions are humanly devised frameworks that shape human interaction (North, 1990), organizations are groups of individuals bound by some common purpose to achieve agreed objectives. A good example to describe the difference between an institution and an organization is to view it as a football game. The 'organization' here constitutes the players and the goal-keeper who take part in the game, while they have to play according to a set of rules and agreements between parties, which is the 'institution'. An organization is viable when it meets the following criteria (adapted from ibid; Debrah and Nederlof, 2002):
- Their members have a common mission or common objective to which they commit themselves.
- All members participate and/or contribute to achieving these objectives.
- The organization functions according to a set of rules (and these are respected).
- The organization mobilizes and manages human and financial resources that allow for enhancing autonomy and sustainability.

Farmers' organizations also respond to these criteria. However, the degree to which they respond may differ substantially and points to the enormous diversity of farmers' organizations. In this bulletin, the distinctive features of farmers' organizations, as compared to other organizations, whether public or private, are:
- Farmers' organizations are rooted in rural areas and related to activities such as primary production, processing and marketing of agricultural products, or related services.
- Members of farmers' organizations strive to improve their conditions (i.e., incomes and well-being) through primary production-related activities; these activities may be subsistence-oriented, market-oriented or a mix of the two.

- Such organizations are membership-based: the organization is led by members and, through collective action, works for its members. They are thus democratic associations of men and women.

A short history of farmers' organizations in Sub-Saharan Africa

Farmers' organizations and groups in Sub-Saharan Africa have existed for a long time, even though they presently occur in forms and structures that are different from before and have evolved in many shapes. The most ancient form of farmers' organization is represented by the 'self-help' groups, where farmers help each other out, for example at peak labour periods and for food security purposes. Such groups are based on social traditions and manage the relations of members within their own local society. These still exist today, but are sometimes overlooked as farmers' organizations, maybe because they are informal and often seasonal (e.g., only during harvest time). Yet they can be important building blocks for networks and genuine, grassroots-based farmers' organizations (see for example the Tanzania cases in Wennink and Heemskerk, 2006).

In colonial times, governments and trading companies introduced certain forms of farmers' organizations to increase their profits, for example to facilitate the production and marketing of export crops. Such organizations were generally 'imported', legal constructions, based on the western cooperative model but managed and controlled by the colonial administration. The French, for example in Benin and Burkina Faso, constituted the *Sociétés Indigènes de Prévoyance* (SIP) later transformed into *Sociétés Mutuelles de Développement Rural* (SMDR) in the former colonies in West Africa (Chauveau, 1992: pp. 2-5). The English in East Africa promoted and facilitated the creation of primary cooperative societies for products such as coffee, tea and tobacco that received support from specialized civil servants. In both cases the main objective was to improve and organize the supply of agricultural products, while linking up with traditional self-help and communal solidarity practices (Ibid). Some of these societies developed into strong, relatively autonomous organizations, e.g., the Victoria Federation of Co-operatives (for cotton) and the Kilimanjaro Native Co-operative Union (for coffee) in Tanzania; this was always due to farmers' demands for stable and acceptable prices, as well as secure markets (Chilongo, 2005).

After independence, many African states (through their newly established government services or 'parastatals')[1] introduced their own types of farmers' organizations, with or without the support of the former rulers, in order to implement state policies (Diagne and Pesche, 1995). In many parts of Eastern Africa (e.g., Zambia and Tanzania), relatively independent cooperative unions were created and managed under government directives, and were later nationalized (Chilongo, 2005). State-controlled farmers' unions were often used to promote cash crop production for export, as an important source of hard currency for the newly independent states. During this period, farmers began to consider these types of farmers' organizations and cooperatives as an extension of the public sector rather than as their own. This explains some of the problems that emerged later in terms of members' affiliation, autonomy, sustainability and ownership of activities undertaken by farmers' organizations (Bosc *et al.*, 2002). Later, many development projects and NGOs also created their own farmers' organizations to constitute an interface between the farmers and themselves, and henceforth facilitate the implementation of the particular activities that they supported. Such projects and organizations often focused on aspects other than specific agricultural products and thus on producers and groups that were not represented in cash crop producer organizations. Besides economic objectives, these other new organizations also had broader community development functions (Diagne and Pesche, 1995).

A great diversity of farmers' organizations

The present situation of a highly diverse picture of farmers' organizations in Sub-Saharan Africa is the result of some recent upheavals, such as the withdrawal of the state from many services, privatization, democratization, liberalization and international dynamics, and the influence of donors on national policy-making (Bosc *et al.*, 2002). As part of these liberalization policies, the state-controlled producer organizations and cooperative unions were reformed, made responsible for their own management and often privatized. Increasing private-sector involvement in the agricultural sector led to the creation of 'outgrowers' associations',[2] often at the initiative of private enterprises. In the mainstream of political democratization, farmers also created their own organizations (e.g., federations, syndicates, etc.) to lobby for and defend their interests at national and provincial levels (see for example the cotton producers' union in Mali; Docking, 2005). In many countries these farmer-led initiatives for new types of farmers' organizations were supported by development cooperation donors and agencies.

The emerging context also shapes the process through which farmers' organizations evolve. More importantly, the context determines the way in which the needs of individuals or households can be fulfilled; either through individual or collective action by joining a farmers' organization (Bosc *et al.*, 2003). The diversity of farmers' organizations is thus explained by several factors such as:
1. origin;
2. legal status;
3. membership base;
4. functions, purposes and services provided; and,
5. scale and level of operations (Ibid).

Origin

This first paragraph of this chapter briefly sketches the history of farmers' organizations in Sub-Saharan Africa. This history already identifies a few possible initiating conditions or establishing entities, such as: a situation where social tradition forms the origin and the organization is set up by farmers themselves to address constraints or exploit opportunities; the state or parastatals; the private sector; NGOs and/or development cooperation agencies; or organizations evolving from farmer groups such as Farmer Field Schools (FFSs), Natural Resource Management Committees (NRMCs) or other 'experiential learning approaches'.

Farmers' organizations can emerge due to farmer-felt needs such as: a need to share local resources (land, labour, water, etc.), market pressures (prices and access to markets), access services (credit, input supply, advisory services, etc.) or for purely social reasons (social security, food security, etc.). In all these cases, there has to be a clear advantage in taking a particular collective action in order to be sustainable; this is often apparent when a need disappears at the end of a particular 'project'.

Legal status

Community-based organizations and common-interest farmer groups can often be either formal or informal, while associations, societies, cooperatives, unions and federations are normally only formal organizations (AgroEco, 2006). Formal groups are registered with the relevant authorities, formed under specific legislation and audited on an annual basis by the government authorities, and under certain conditions, governments can cancel the registration. Formal organizations, particularly the larger ones, have a professional management team, whilst this is lacking in most informal groups. Larger formal groups generally engage in structured activities related to their objectives and create by-laws or a constitution, whilst informal groups can often be more flexible and engage in unstructured self-help activities, without a (written or verbal) code of conduct. Formal groups often belong to a local, national or international network, whilst networks amongst informal groups are limited (Ibid).

Membership base

Farmers often organize (or are being organized) according to the commercial commodities they produce: e.g., coffee, rice, cotton, cashew or cocoa. Such organizations usually group large-scale, agribusiness-like farms with commodity-oriented smallholders. 'Family farms' form a very large group among the members of farmers' organizations because this is the way agriculture is generally organized in Africa: there is a strong link between economic activities and the family structure, its wealth and labour resources (Bosc *et al.*, 2003). However, for many farm families it does not make sense to focus on only one crop or dimension of their enterprise. It is the combination of different crops and key strategies that explains the complexity of their farming system and groups can be organized accordingly.

Functions, purposes and services provided

Another way often used to distinguish between farmers' organizations is according to their functions, purposes and the related services provided. The simplest defines three categories of functions:
1. service provision;
2. advocacy and lobbying; and,
3. communication and coordination (adapted from Collion and Rondot, 1998).

However, some farmers' organizations take a more activist and political position, as has often been the case in Latin America and also in pre-independence Africa (Bebbington and Thompson, 2004: Chilongo, 2005).

Scale and level of operations

Farmers' organizations can link and unite at levels other than local ones, and can form unions, federations, networks etc. Two pathways for farmers' organizations to unite are encountered most in Sub-Saharan Africa. In the first scenario, farmers' organizations integrate at different levels around a given commodity (e.g., cotton) with specialized functions and services at each level. The local level handles the logistics for input supply and product marketing; the provincial level provides technical and management support to the local groups; and the national level is involved in policy-making and negotiations about the enabling environment, such as price setting for inputs and products, as well as government taxes and subsidies. The processes and approaches followed have often been encouraged by governments and donors as part of the privatization process and withdrawal by the state from supporting functions. A second trajectory is the one followed by federations, networks etc. that are successful in defending the farmers' causes and mobilizing resources for projects. Their successes attract organizations that want to become affiliated in order to gain perceived benefits (Bosc *et al.*, 2002).

Typology of African farmers' organizations

Farmers' organizations can be classified into groups that may eventually provide a basis for a typology. Classification is a means to distinguish and describe different farmers' organizations with the aid of one or more criteria, for example female, male and mixed organizations, or managing a cereal bank, irrigation scheme etc.

A typology goes one step further and aims to analyze the dynamics of farmers' organizations as organized entities within a given context, which subsequently allows for designing strategies for further intervention. Typologies are meant to accompany processes, and hence a typology is not a 'fixed state' but an instrument with which to understand and analyze organizations, for example when designing support programmes (Pesche, 2001).[3]

Common criteria for classifying and/or establishing a typology of farmers' organizations are related to the factors discussed above (for a summary see Table 3). The growing attention and interest in farmers' organizations over the last decade has also led us to approach them from a perspective of institutional development and organizational strengthening, with tools that are being used in civil society or the third sector (e.g., NGOs and community-based organizations). For example, assessment tools allow for monitoring capacity-strengthening trajectories and the development of social capital (Gubbels and Koss, 2000). The priorities defined in these areas, whether they were explicit or implicit, have consequences for the future position of farmers' organizations. For example, human resource development in village cooperatives for developing commodity sectors (coffee, cacao or cotton) by parastatals, was mainly aimed at improving the logistics for input supply and providing a reliable supply of products. This largely explains their current focus and ties with the private sector, as well as the social capital that they have built up (Bingen *et al.*, 2003).

Although these criteria (Table 3) are useful when aiming to elaborate a typology of farmers' organizations, one has to take certain precautions:
- Each of the criteria is just one facet of an organization's identity, which in turn reflects the society and livelihoods of its members. Several criteria need to be used together in order to grasp the complexity of farmers' organizations.
- A 'simple' application of the criteria produces a rather static picture of farmers' organizations. Its evolution and dynamics as an organizational entity, within a given context and compared to similar organizations, are much more interesting. For this purpose, a 'scale of values' for the various variables may help to comprehend the development trajectories of farmers' organizations (e.g., homogenization or diversification of the membership; specializing or generalizing through its functions; scaling-up or scaling-down of operations).

Farmers' organizations as interfaces

Today's farmers' organizations in Sub-Saharan Africa are often hybrid organizations (and difficult to distinguish from NGOs) through their variety in status, missions, membership-bases and financial sources for functioning. So, numerous farmers are currently members of more than one farmers' organization in order to have, through

Table 3: Examples of the most common criteria for classifying farmers' organizations

Criteria	Variables
Origin	As an autonomous organization in reaction to constraints or opportunities; emerging from the local community.
	As an organization created by outside interventions: (a) the state or parastatals; (b) the private sector; and/or (c) NGOs and development cooperation agencies.
Formal and legal status	Not registered with the relevant authorities.
	Registered under various legislation and facilitated by the relevant authorities: (a) 'association' with the Ministry of Home Affairs; (b) 'cooperative' with the Ministry of Agriculture or cooperative organizations; or (c) 'union' with the Ministry of Labour.
Membership base	On a sub-national basis; related to an administrative entity.
	On the basis of farm size and market orientation: (a) large-scale, agri-business farmers; (b) small-scale, commodity farmers; and/or (c) subsistence-oriented, family farms.
	On the basis of farming systems: (a) agriculturalists; and/or (b) livestock keepers; or (c) mixed farming.
	On the basis of social groups (i.e., gender): (a) one specific group; or (b) a mix of groups.
Functions, purposes and services provided	Functions: (a) economic; (b) social; (c) representation, such as defending interests, lobbying and advocacy; (d) communication, sharing of information and capacity building; and/or (e) coordination.
	Purpose: (a) single purpose, specialized in one commodity, activity or sector; or (b) multi-purpose.
	Services provided to members: (a) input supply; (b) marketing of products; (c) access to new technologies; and/or (d) technical and management training.
Scale and level of operations	Levels: (a) village/district; (b) province; (c) national; and/or (d) international.
Organizational structuring, governance and management procedures	Very few, or no, organizational structures and/or documented procedures for governance and management.
	Emergence of functioning, organizational structures and respected, documented procedures to enhance good governance and management.
	Complete organizational functioning, with a set of documented procedures that are being respected.

Sources: adapted from Beaudoux and Nieuwkerk, 1985; Bebbington and Thompson, 2004; Bosc *et al.*, 2003; Gubbels and Koss, 2000; Pesche, 2001; and Messer and Townsley, 2003.

collective action, access to resources and services provided by the organizations or third parties.

Another, more practical reason for such multi-membership is that many farmers cultivate more than one crop, whereas farmers' organizations often focus on only one commodity. In other cases, farmer leaders have gained legitimacy towards other organizations and, more importantly, have developed networks and skills to mobilize resources, for example from development cooperation agencies. Their organizations have often become successful intermediaries between farmers and other stakeholders in the development cooperation sector. Farmers' organizations therefore operate as the interface between the farmers at village level and their overall environment (Roesch, 2004: see Figure 2).

Figure 2: Farmers' organizations at the interface between local and global society

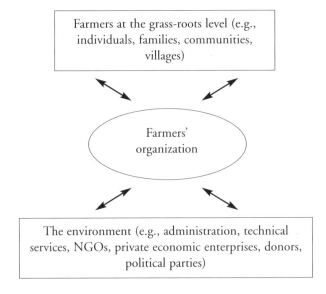

If an external stakeholder measures the efficacy and performance of a farmers' organization, an important criterion will be the degree to which the organization can initiate change at the grass-roots level (Ibid). When the farmers' organization is created by the local society, the organization generally aims at serving to influence its environment and often constitutes a negotiation force.

However, in many cases the external actors are dominant and farmers' organizations are encouraged to adapt to their environment. Examples are the many 'learning platforms' that have emerged, such as groups that develop a technology together or learn about crop and pest management (e.g., FFSs) or other platforms for experiential learning. Under the influence of outside partners, when attempting to institutionalize or scale-out their approach, such groups often transit into more structuralized entities (Gallagher, 2001).

It is often the role of the group leader to find a balance between adapting to the environment and satisfying the needs of members, while keeping in mind sustainability (in terms of genuine grass-roots support and access to financial resources). Farmers' organizations are continuously adapting because:
1. they have to adjust to the environment;
2. their role at grass-roots level changes; and,
3. the roles of the farmers within their organizations change (Roesch, 2004).

This also explains why farmers' organizations in Sub-Saharan Africa often fulfil multiple functions and pursue several purposes at the same time, which seem difficult to combine. This means that perceptions by the stakeholders involved need to be taken into account, since farmers' organizations present different goals and means for different stakeholders (Chirwa et al., 2005).

Notes

1. Parastatal companies are enterprises or organizations that are wholly or partially owned by the state. Although they may have a certain autonomy in management, the government defines the composition of the supervisory board and policy guidelines.
2. Outgrowing is a form of contract farming: farmers produce certain products on their own land under a contract with a processor or trader who guarantees the purchase of these products, which have to meet predefined standards.
3. In general, a farmers' organization can play more than one role for its members. Pesche (2001) stresses that roles or functions are probably not a useful base for a typology because an organization's activities are just one facet of its identity.

3 Analytical framework

Analyzing the role of farmers' organizations in enhancing social inclusion in the access to agricultural services requires an understanding of the way in which farmers are organized, and the mechanisms for social exclusion of agricultural services provision. In this context social inclusion hindrances refer to:
1. lack of assets (resources, social relations etc.); and
2. institutionalized barriers to access services (see Chapter 1).

Social inclusion hindrances may restrict farmers from becoming members of an organization and hence having less access to services. For example, Silver (2004) admits that the Hoima District Farmers' Association (in the north of Togo) does not work with the 'poorest of the poor', even though his organization intends to do so. The farmers he works with are the ones who can pay their annual subscription fees: 'The farmers we deal with can afford to buy bicycles, radios, have semi-permanent and permanent houses, and some are market-oriented, thus producing targeting the markets. They produce from an average minimum acreage of half a hectare, which are not the very poor in Togo'. However, it is often the Sub-Saharan African women who are 'the poorest of the poor' and these are excluded. Although they play an important role in agriculture, their role is not always fully acknowledged. In addition, they are often submitted to 'traditional' institutions (e.g., power relations and land tenure), which may lead to their marginalization within rural society, with no access to services and no opportunity to join farmers' organizations (FAO, 2007).

These same hindrances may hamper certain groups to fully exercise their rights as members, by freely expressing their needs, being elected as a leader or accessing services provided by the organization or third parties. Examples of such groups include groups of women, farmers without their own land, distressed households (HIV/AIDS, malaria etc.) or ethnic minorities. Criteria and rules, whether formal or informal, may exclude certain member groups from being represented or fully participating in a farmer's associative, democratic life. The composition of the governing and administrative bodies, and the mechanisms involved, are also a reflection of the role of farmer members within the organization. Furthermore, communication between members, leaders and staff (particularly member participation in policy and strategic decision-making) may be more difficult in larger organizations with several organizational tiers. The representative function of a farmers' organization, to which many service providers refer when seeking to collaborate for reasons of effectiveness and efficiency (i.e., 'economy of scale'), also raises the question of whether the organization represents only farmer members or also non-members within the sector or area.

As previously mentioned, the main drive for farmers to organize themselves is that collective action, rather than individual action, provides a better opportunity to gain a suitable response to their needs (Bosc *et al.*, 2003). Trust, reciprocity, cooperation and communication are therefore crucial, since they allow for collective action and lowering of 'transaction costs' in situations where formal contract development and enforcement is difficult (Grootaert and Van Bastelaer, 2002). The ties within a farmer group ('bonding social capital' of associations, cooperatives etc.) may be enhanced beyond a given group and may include other farmer groups ('bridging social capital' of unions, federations etc.) to develop collective action at other levels or in other areas. Finally, farmers' organizations may develop relationships with government authorities, as well as public and private service providers, in order to influence decision-making towards the well-being of their members ('linking social capital'). Figure 3 presents the different forms of social capital. This is why social capital is considered a crucial asset in improving the livelihood system and hence to overcome social exclusion in access to services. However, strong social capital is not a guarantee of social inclusion, since norms within an organization may still hamper certain groups (such as women farmers or minority ethnic groups) from accessing services (Ibid.).

The various elements that have been mentioned above (membership, gender, representativeness and farmers' roles, and social capital) affect the role of farmers' organizations in service provision. Services, including capacity strengthening, can be; either self-provided or provided by third parties; and either to their members only or to the community as a whole. Yet, the inclusive character of services also depends on the service providers themselves and is often the result of a continuous interaction (see Table 2). The role of farmers' organizations in this interaction can be threefold, i.e., to:
1. persuade services to listen to the poor and vulnerable among their members and non-members, and facilitate the voicing of these groups;
2. influence the agenda of services (e.g., setting priorities for research and extension); and,
3. provide and supply these services on a joint basis or by themselves.

Figure 3: Bonding, bridging and linking social capital

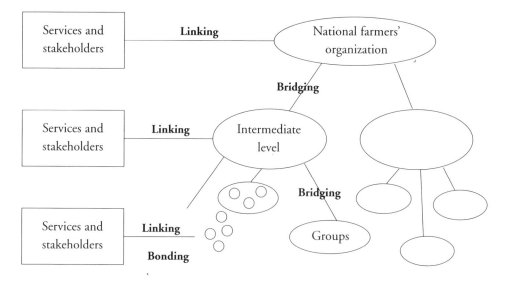

The general considerations listed above lead up to an analytical framework based around six main topics:
1. membership of the farmers' organization;
2. gender;
3. social capital;
4. representativeness of the organization;
5. the role of farmers within the organization; and
6. the role of farmers' organizations in accessing service provision.

Table 4 shows the key issues for each of these topics. This Table also presents the dimensions of social inclusion (see Table 1) and the links with criteria for typifying farmers' organizations (see Table 3). Knowledge of the main characteristics of a farmers' organization, according to these criteria, is considered to be a prerequisite for understanding their role in enhancing socially inclusive service provision.

The results from applying the framework to data and information gathered through case studies permits:
- Identification of internal factors (organizational weaknesses and strengths) and external factors (context-related opportunities and threats) that influence social exclusion and inclusion.
- Assessment of the role played by farmers' organizations in accessing services for their members, non-members and the poorest among them.
- The proposal of changes for facilitating access to services for the poor.

Table 4: An analytical framework for the role of farmers' organizations (FO) in enhancing socially inclusive service provision

Criteria for typifying an FO[a]	Topics	Issues for social inclusion	Dimensions[b]
Origin Membership base	Membership	Number of members Activities of members Criteria (formal and informal) to become a member Costs and benefits of membership	Cultural Political
	Gender	Mechanisms for excluding female members Strategies for including female members	
Purposes, functions and services Scale and level of operations Formal and legal status	Role of the FO in service provision	Role of the FO in accessing services Role of the FO in enhancing access to services for the poorest (members and non-members) Strategies of service providers for enhancing access for the poorest	Economic Functional Physical Relational Structural
	Social capital	Bonding and bridging capital Linking capital	
Organizational structuring, governance and management procedures	Role of farmer members	Composition of governing bodies Mechanisms for constituting governing bodies Role of the poorest in leadership	Participatory
	Representativeness	Meaning of the FO for non-members Downward and upward links	

[a] The main criteria that are relevant (see Table 3).
[b] The main dimensions of the 'inclusion lens' concerned (see Table 1).

4 Methodology of case studies

The comparative analysis of the case studies aims to:
1. investigate the current role of farmers' organizations in facilitating access of the poorest to agricultural services; and
2. identify the conditions under which farmers' organizations can enhance social inclusion.

Farmers' organizations were selected according to the following criteria:
1. origin and membership base, to include both commodity-based organizations and network-based organizations;
2. originating from countries in both Eastern and Western Africa; and,
3. having a 'partnership' with KIT (this allowed for more easily and rapid access to information sources, also involving leaders and members in debates concerning social inclusion).

The cases concern:
1. KILICAFE (the Association of Kilimanjaro Specialty Coffee Growers) in Tanzania;
2. UCPC (three District Unions of Cotton Producers) in Benin;
3. ACooBéPA (the Association of Cashew Growers' Cooperatives in Benin);
4. MVIWATA (the Network of Tanzanian Farmer Groups) in Tanzania;
5. INGABO (the Union of Farmers and Livestock Keepers) in Rwanda.

Three of the five cases concern organizations that were also involved in earlier case studies and action research (see Heemskerk and Wennink, 2005; Wennink and Heemskerk, 2006).

The cases studies were conducted in 2005 by staff members from the farmers' organizations involved (Tanzania), from partner organizations of the farmers' organization (Rwanda), or by associated researchers (Benin). KIT researchers elaborated the terms of reference and checklists for the case studies and provided feedback during the fieldwork and reporting. The first case study results were presented during workshops with leaders and members (Tanzania and Rwanda) or submitted as reports to leaders for feedback (Benin).

In all cases, several methods of collecting data and information were employed, such as: desk study of policy documents, membership records, and activity reports; semi-structured interviews with members and leaders; and focus group discussions with small groups of members, leaders and staff. The researchers ensured triangulation of methods to obtain the same information or contact stakeholders, to check whether different stakeholders had the same, or differing opinions, on issues. However, qualitative data rather than quantitative date was gathered during the case studies, and gaps were filled through additional desk studies by all researchers involved.

The results presented in the case study reports were analyzed according to the framework shown in Table 4. This allowed researchers to compare the different situations and draw conclusions on the role of farmers' organizations and social inclusion in providing agricultural services (see Chapter 5).

Each case study (see Part II) presents information on:
- The situation in the country with regard to farmers' organizations, the overall policy and institutional context, and agricultural service providers; the farmers' organization, its origin and basic characteristics, such as status, organizational set up and areas of intervention.
- The main elements of social inclusion (membership base, gender, social capital, representation and participation of farmer members, plus the role of the organization in accessing services for the poor); the key issues involved; plus the authors' conclusions on the level of inclusion of the farmers' organizations in relation to access to services by the poorest.

5 Farmers' organizations and social inclusive service provision

Types of farmers' organizations

This bulletin focuses on the role that farmers' organizations play in enhancing the poorest farmers' access to services such as research, training, advice and extension. We assume that greater access by the poorest farmers to services improves their innovative capacities and thus their livelihoods. However, before we can engage in this discussion we need to know what types of farmers' organizations were studied in the cases selected (see Part II). An overview of the characteristics of each farmers' organization studied is shown in Tables 5a, 5b and 5c.

The cases studied concern three commodity-based organizations (i.e., cashew, coffee and cotton) and two more general networks for farmer groups. However, one of these is also actively involved in organizing members among cash crop farmers, as a way to improve members' incomes and reinforce the resource base of the network (i.e., INGABO, which helps to organize cassava growers).

In the case of commodity organizations, the organizational levels follow the logic of the supply chain, whereas networks seem to follow the formal administrative entities within a country. All the organizations received outside support to help them get established. However, in some cases the initiative clearly came from the donor (e.g., ACooBéPA) whereas, in other cases, external agencies provided support (e.g., MVIWATA, which originally emerged in collaboration with the local university; and INGABO, which is a member of a network of farmers' organizations and national NGOs). Therefore INGABO, KILICAFE and MVIWATA can be typified as farmer-led movements, whilst ACooBéPA and UCPC seem more 'outside' initiatives, with support from donor-funded projects (ACooBéPA) or state services (UCPC). This difference in establishment has a huge impact on the degree of ownership amongst beneficiaries.

It goes without saying that laws and regulations on legal status have an enormous effect on the potential impact of farmers' organizations. In some cases each level of the organization has a different legal status. So, local level groups can be either associations or cooperatives with a fairly informal nature and anchored in more traditional community-based organizations (e.g., INGABO). National and sub-national farmers' organizations are often registered under specific legislation, facilitated by the Ministry of Agriculture, Cooperatives, Labour or Commerce (e.g., KILICAFE, as a limited company, and INGABO as a union). In many cases such legislation has not been adapted to encompass the specific situations and requirements of farmers' organizations.

With respect to the mission statements of the farmers' organizations, we note that few explicitly mention poverty alleviation goals or inclusion issues. This is particularly the case for commodity-based organizations: access to markets and a reasonable share of market prices are considered crucial to improving the income of all members. Crop growing, as a membership criterion, without any distinction between farm sizes, is seen as a prerequisite for equal access by members to services provided. The network organizations generally focus explicitly on smallholders.

With the exception of ACooBéPA, which covers two districts, all other farmers' organizations have quite a large coverage area and consequently have several organizational layers while also operating at these levels. Organizational functioning is clearly a question of time (age) and experience in developing and establishing the necessary structures and procedures at the different levels. Voicing grass roots opinions remains a continuous challenge. The functionality of farmers' organizations depends on the level in their hierarchy: for example, organizations participate in price negotiations at the higher levels, while local level groups are concerned with input distribution and the sale of produce. Both INGABO and MVIWATA strive to strengthen service provision at both higher and lower levels (grass roots and members) by developing and concluding 'service contracts'.

Farmers join farmers' organizations for a variety of reasons: whereas farmers often use commodity-based organizations to gain access to markets, inputs or credit facilities, networks are also considered important for general countervailing concerning service providers and in order to gain political clout. We observe that, in addition to the public sector (which has traditionally provided extension services), many farmers' organizations have become increasingly involved in providing technical advice. However, this is not the case in Benin, where the public extension service continues to be paid through cotton levies and hence farmers' organizations do not see the need to also intervene themselves. Also, all farmers' organizations have an increasingly strong economic dimension and market orientation because, in general, national agricultural development policies emphasize value-chain development and market access.

Membership

The preceding chapter discussed the types of farmers' organizations that were analyzed, including their focus (e.g., crops or farming/livestock keeping). Tables 6a and 6b provide an overview of membership characteristics per farmers' organizations studied.

In the case of ACooBéPA, UCPC and KILICAFE, specific groups of farmers (growing a particular crop) can become members. In the case of INGABO, individuals (farmers) are members of the organization and, in turn, these members can also be members of a local farmer group, but this is not a condition for membership of INGABO. In the case of MVIWATA, individuals who are members of a farmer group, the farmer groups as a whole, as well as complete networks of farmers can all become members. It can therefore be concluded that the commodity-based organizations represent groups, whereas the network organizations focus on advocacy and lobbying for more general farmer-related issues, and also represent individual members. The latter type of organization is possibly linked to the more syndicate-like lobbying and advocacy characteristics.

In most cases landownership is a de facto condition for membership, since farming as a means of living or growing a specific crop is the basic criterion, even though in some cases land-user rights are enough. In Benin for example, farmers can be land users and still belong to the village groups that are UCPC members. In Rwanda the criterion of earning most of the household income from farming, in combination with the high land pressure and the importance of off-farm activities, mean that it is usually the landowners who are members of INGABO. Only in the case of MVIWATA are landless farmers specifically mentioned as becoming full-fledged members, if agricultural production is the main component of their livelihoods.

In general, it seems, farmers' organizations do not have a very clear picture of their members (in terms of land property, age, education, religion, ethnicity etc.). There is some knowledge of diversity, but this us not used for targeting and improving service provision to members. It is therefore recommended that farmers' organizations develop such a profile, since this would help to take membership diversity into account. This could also help to verify the extent to which the members of the organization are representative of the entire community.

In all cases members are required to pay entry fees as well as annual fees. The ACooBéPA and UCPC, and their affiliated village groups, also require a social share to constitute a working capital for the cooperatives. Regular payment of membership fees is a recurrent problem and it remains to be seen whether this is due to lack of financial means (e.g., the low incomes of members or financial mismanagement) or

Table 5a: Characteristics of the farmers' organizations (FO) investigated in the case studies

Type of FO	ACooBéPA Benin	UCPC Benin	INGABO Rwanda	KILICAFE Tanzania	MVIWATA Tanzania
Origin	Public-private initiative (project) with donor support	Farmer-led, with public sector support (for the cotton parastatal)	Farmer-led, with donor and NGO support	Farmer-led, with private sector and NGO support	Farmer-led, with external (university) support
Trigger	National policy on crop diversification (projects)	Cotton marketing: sector-wide reforms and efforts toward privatization	Lack of infrastructure and services (due to genocide)	Coffee marketing: auction thresholds	Loss of trust in old Cooperative Unions
Legal status	District and local levels: not yet registered as Cooperatives	District and local levels: registered as Cooperatives	Provincial/national level: registered as Union Local level: traditional associations that are not formally registered	National level: registered as Limited Company Intermediate level: legally as branch offices of KILICAFE Local: FBGs as cooperatives or associations	National level: registered as Trust Fund Intermediate level: networks as NGOs Local: groups as cooperatives or associations
Base	Cashew	Cotton	Farming/Livestock keeping (smallholdings)	Arabica coffee	Farming/smallholder farming
Purpose	Improved prices and service provision	Representation and coordination	Representation and advocacy at national level Service provision	Service provision and improved prices Service provision at local level	Representation and advocacy at national level Service provision at lower levels

44

Table 5b: Characteristics of farmers' organizations (FO) investigated in the case studies

Type of FO	ACooBéPA Benin	UCPC Benin	INGABO Rwanda	KILICAFE Tanzania	MVIWATA Tanzania
General services	Access to inputs (seeds) and marketing Training and advice for improved production	Access to inputs and marketing Training and advice for improved production Participation in community development	Training and extension services for crop and animal production Access to credit Management support for specialised producer federations	Improved coffee quality and processing Access to markets and credit	Advisory services on lower levels Access to markets and credit
Scale of operation	In 2 out of 112 districts in Benin; 10 wards and 36 villages Heart of the cashew area	In 75 out of 112 districts in Benin. Quasi coverage of cotton area	Member groups in one province of Rwanda; in 800 out of 1,097 cells	In 5 out of 25 provinces in Tanzania: 3 medium-sized networks and 102 FBGs	In all 25 provinces of Tanzania; 150 local and medium-sized networks in 80 out of 114 districts; 1,100 FGs
Level of operation	Villages, ward and district	Villages and district Member of provincial and national unions (FUPRO)	Village, district and province Member of a national network (ROPARWA)	Village and intermediate level	Villages, local, intermediate and national
Organizational structuring and governance	Organizational structures defined but not properly functioning Grass roots = weak No regular AGM Assets limited	Organizational structures defined and functioning Strong AGM at district level Grass roots = strong Assets limited	Organizational structures defined and properly functioning Strong AGM at district level Grass roots = strong Assets limited	Organizational structures defined and properly functioning Strong AGM at national and intermediate (chapter) level Grass roots = strong Assets at all levels	Strong AGM at national and intermediate level Grass roots = weak Assets limited

Table 5c: Characteristics of the farmers' organizations (FO) investigated in the case studies

Type of FO	ACooBéPA Benin	UCPC Benin	INGABO Rwanda	KILICAFE Tanzania	MVIWATA Tanzania
Documented procedures	Not available	Statutes and by-laws, but not up to date Management and financial manuals No MoU	Statutes and by-laws Management and financial manuals MoU (service contracts) under preparation	KILICAFE and FBG Constitution linked through MoU	Constitution Management and financial manuals No MoU

Table 6a: Characteristics of farmers' organization membership

Membership	ACooBéPA Benin	UCPC Benin	INGABO Rwanda	KILICAFE Tanzania	MVIWATA Tanzania
Type of members	Smallholder cashew growers (groups)	Smallholder and large holder cotton growers (groups)	Smallholder farmers and livestock holders	Smallholder coffee growers (1-2 hectares)	Smallholder farmers and fishermen
Number of members	546 individual members through 36 village groups and 10 ward associations	Numbers not available Most households (>90%) are members in the cotton area (Kandi) In the other two areas: a minority of households (<50%)	13,000 individual members through 800 groups (an estimated 50% of households)	10,000 members (of 400,000 households producing either Arabica or Robusta)	10,000 individual card holders and 60,000 members through 1,100 groups and 150 networks

Table 6b: Characteristics of farmers' organization membership

Membership	ACooBéPA Benin	UCPC Benin	INGABO Rwanda	KILICAFE Tanzania	MVIWATA Tanzania
Formal criteria to become a member	Cashew grower (village group member)	Cotton grower (village group member)	Farming/Livestock keeping Agriculture as main source of income	Coffee producer (FBG member)	Group member and farming as main livelihood
Informal criteria to become a member	Cashew plantation owner	Village groups need to produce 'enough' cotton	Landowners	FBGs need to produce minimum 10 MT coffee	No civil servants or politicians
Costs of membership	Levies on sales (not functional) Social shares (USD 5), annual fees (USD 2) for combined individual membership of village and ward groups	Levies on cotton sales Social shares, entry and annual fees (variable) for individual membership of village groups and village group membership of the union	Entry (USD 2) and annual fees (USD 2) for individuals Contribution to a village group credit and mutual fund (USD 0.5 per member per month)	Payment: 3% of gross coffee sales, USD 25 entry and USD 25 annual fee for FBGs, and USD 2-4 for members	Entry fee and annual fee for individuals (USD 1 and USD 1), groups (USD 2 and USD 10) and networks (annual USD 20)
Benefits of membership (potential)	Village group members: access to inputs (seeds) and markets, and research and extension services	Village group members: access to inputs and markets Representation with local service providers Advocacy and lobbying (FUPRO)	Village group members: access to credit fund and extension services Representation and advocacy with service providers Mobilization of donor support for group activities	Coffee marketing (premium price), and access to FBG assets (processing units) and credit	Group benefits are variable but access to markets and credit Representation, lobbying and advocacy AGM member

other reasons, such as weakly perceived benefits from services provided by the organization to its members. In the commodity-based organizations (such as KILICAFE and UCPC), levies (and rebates) provide revenues to run the farmers' organizations. In the case of UCPC, these levies are destined to reward farmers' organizations for cotton chain operations such as the collection of cotton for marketing. In the case of cotton, the funds generated through levies are influenced by the performance of the entire cotton chain and are therefore highly dependent on world market prices. These are two issues on which the UCPCs apparently lack intelligence and information, as well as the capacity, to intervene.

Both network organizations, MVIWATA and INGABO receive significant amounts of donor funding for various project-related activities. Such donors often aim to alleviate poverty through the projects that they support, and the network organizations (rather than the commodity-based organizations) are therefore more likely to design interventions to fight poverty and enhance social inclusion. Obviously this makes the networks quite vulnerable to donor's conditions and might present a threat to the long-term financial sustainability of the network. INGABO is therefore considering a proposal to split the ROPARWA network into two entities: one to manage donor funds and project implementation, and another to concentrate on advocacy and policy-making. MVIWATA has established a trust fund for several reasons, including to maintain its assets and to become less dependent of external support. However, donor funds, which are often allocated to support poverty-alleviation activities, might also represent an opportunity to address the specific needs of the poorest rural people.

Gender

Social exclusion can be experienced in gendered ways, and gender can ameliorate or exacerbate exclusion or the terms of inclusion (see Beall and Piron, 2005: p. 22). Table 7 shows details of female/male membership levels. For another category of members, the young, unfortunately no data on characteristics and circumstances of members was available, since this did not seem to be a priority among the organizations involved.

As a general trend, fewer women are represented in 'classical' commodity organizations. This is closely linked to the position of women in Sub-Saharan African rural society and their access to, and control of, production factors such as land. The UCPCs in Benin, for example, have hardly any women (less than 30%) who are registered members. Women are not officially refused membership, but informal criteria (such as the quantity of cotton produced) often exclude women, since they usually have the smallest plots. Literacy is also often an asset for those occupying leadership positions, and women are generally less literate. Those networks that increasingly focus on market access and value-chain development accordingly have a larger percentage of male members.

However, the recent feminization of the agricultural sector (as a result of the HIV-AIDS pandemic and mass exodus searching for labour) has resulted in an increased focus on women. Hence, many organizations now have a set of specific gender-facilitating policies to increase the proportion of women members. On the one hand, this is the result of the aforementioned feminization of the agriculture sector, which underlines the important role played by women, and allows them to actively participate in managing organizations and to voice their needs. On the other hand, donor-supported initiatives to develop skills help women to claim their rights within organizations. In conclusion, commodity-based organizations tend to be rather 'gender-blind' (e.g., KILICAFE) or at least 'gender-unaware' (e.g., UCPC and ACooBéPA), while network organizations are even 'gender-distributive' (e.g., INGABO and MVIWATA) (see MacDonald et al., 1997: p. 52 for models of gender and organizational change). The two network organizations clearly took up the challenge to include women in both membership and leadership roles as a result of developments within rural society, as well as gender sensitization by donors. However, these organizations have apparently taken care to do so at their own pace, and thereby ensure 'ownership' of the issue and to institutionalize gender-specific measures.

Participation and representation of farmers

Once farmers are members of a farmer organization, important issues are the division of functions between members and leaders, as well as the representative character of the organization for different groups of farmers, particular the poorest. Tables 8a and 8b present an overview of these issues for the farmers' organizations that have been studied.

In all commodity organizations large-scale farmers and former (now retired) civil servants have the advantage in gaining leadership positions. This is because, on the one hand, these farmers are motivated to defend the interests of other farmers and are aware of some of the mechanisms that can be used to defend farmers' rights, while on the other hand they can afford to spend some time and money on group interests. Furthermore, the levy system for funding organizational functioning favours large-scale farmers, since they claim more influence because of their financial contribution (e.g., the election of UCPC board members). In the network cases, small-scale farmers have more chances of gaining leadership positions, certainly at the local level – however, it is not clear how often this actually happens. Other important eligibility criteria for leadership positions include communication skills and the candidates' social capital, including links with the local elite and political system. Poorer members are less likely to occupy leadership positions because they are less likely to

Table 7: Gender dimension of farmers' organization membership

Gender	ACooBéPA Benin	UCPC Benin	INGABO Rwanda	KILICAFE Tanzania	MVIWATA Tanzania
Inclusion/exclusion mechanisms for female members	20% female members of village groups. Few women in leadership. No gender strategy	20% female members of village groups. Few women in leadership. No operational gender strategy	60% female members. Women in leadership. Gender quota strategy	10% female FBG members. Few in leadership. No gender strategy	18% female members (was 30%; FGs with higher percentage) 44% in national leadership. Gender strategy
Inclusion/exclusion mechanisms for young members	No data	No data	No data	No disaggregated data by age. FBG members are said to be relatively old	No disaggregated data by age

Table 8a: Role of farmers within the farmers' organization (FO): participation and representation

Farmers role in FO	ACooBéPA Benin	UCPC Benin	INGABO Rwanda	KILICAFE Tanzania	MVIWATA Tanzania
Office composition	Village group committees: smallholders. District board: smallholders	Village group committees: smallholders and large-scale farmers. District board: large-scale farmers	Village group committees: smallholders. Central committee: smallholders and large-scale farmers	FBG management committees, Chapter Committee, and Central Management Team: all smallholders	Smallholders dominate at local level (44% women) Secretariat with appointed staff (only 22% women)
Mechanisms for constituting an office	District Board is elected by AGM	District Board is elected by AGM, and appoints executive board. Cotton production and residency weigh in voting	Board of Directors is elected by AGM, and appoints executive directors	Governing Board is elected by AGM, and appoints executive secretariat	Steering Committee is elected by AGM, and appoints executive secretariat. Politicians excluded

Table 8b: Role of farmers within the farmers' organization (FO): participation and representation

Farmers role in FO	ACooBéPA Benin	UCPC Benin	INGABO Rwanda	KILICAFE Tanzania	MVIWATA Tanzania
Role of the poorest in leadership	Moderate Tendency toward dominance of major ethnic groups	Limited because of (informal) election criteria	Limited, as large-scale farmers are better endowed (for elections) Institutionalized gender strategy led to improvement in leadership roles No landless farmers	Limited, as few households and coffee producers are not the poorest	Few landless farmers Gender strategy led to improvement in leadership roles
FO's meaning to non-members	FO involved in extension services (project-funded and NGO-managed), which are accessible to on-members	Financing of community infrastructure Funding of extension services (cotton levies), also for non-members	Representation, lobbying and advocacy with results (e.g. laws, infrastructure), also for the benefit of non-members	FO can buy coffee from non-members	FO involved in extension, also for non-members
Downward links	Loosely accountable to members (no regular AGM)	Through AGM and rural radio	Through AGM	Strongly accountable to FBGs	Only through AGM
Upward links	Weak because of 'shielding' by project partners	Strong, since cotton is important cash crop	Strong, through networking with other FOs	Accountability of FBGs on coffee quality	Weak presentation of innovation demands by groups
Access to services	Agenda setting of research and extension services by donor-funded projects and NGOs	Funding of agricultural extension and cotton research (cotton levies), and involved in priority setting	Joint activities with agricultural service providers (donor-funded and NGOs)	Member of Tanzania Coffee Research Institute and influences its agenda, coffee processing and quality services	Strong in setting research priorities at national level, although variable Involved in agricultural extension services

be literate, less likely to have time available and generally possess less social capital. In addition, they are not likely to be considered role models (e.g., successful agricultural entrepreneurs who build up social capital) by their peers.

In some cases leadership is hierarchical: a member has to obtain a leadership position at a lower level before becoming eligible to a leadership position at a higher level (i.e., a 'ladder system'). This is the case for the UCPC and ACooBéPA in Benin and INGABO in Rwanda. On the contrary, KILICAFE and MVIWATA encourage separate leadership, so that anyone can aspire to a leadership position. The advantages of a 'ladder system' mean that representativeness is legitimized, and anchored at lower levels of the organization, and leadership capacities are optimized. Ordinary members can also consult 'their own leaders' and, through them, reach the higher layers. However, it also results in a trade-off between quality and transparency. Advantages of an autonomous leadership system include the high degree of transparency, a broader base, capacity building of more individuals and improved accountability. In all cases there is a tendency to form elite leaders, which in itself is not a problem as long as they account for their decisions and members have the countervailing power to control them. Failure of such democratic mechanisms put the social cohesion of the farmers' organization at stake, and annihilates motivation for collective action.

In both the INGABO and MVIWATA cases (organizations originally meant to be for the benefit of all farmers, including the poorest) an increasing focus on marketable commodities is noticed, which may have consequences for participation and representation. In the MVIWATA case it has already been observed that the number of female cardholders is still growing, but not as fast as the overall membership, and thus the proportion of women members is decreasing, which seems to be linked to the focus on marketable commodities.

With respect to the representativeness of the farmers' organizations for non-members, network organizations tend to initiate activities that are beneficial for non-members and include consultations with non-members, through their focus on lobbying, advocacy and collaboration with public-sector service providers and NGOs. Economic services, such as input supply and marketing of products that are provided by commodity organizations, are exclusively for members. However, research and extension services that are undertaken jointly and/or funded by commodity organizations are also accessible to non-members.

Downward links, where the national organizations consult their membership base to guide decision-making, become rarer as the size of the organization increases. This can easily be explained by the size of the group and the distance (also physical) between the different layers within the organization. However, regular consulting is crucial for transparency and accountability within the organization and its social capital. Lack of transparency creates mistrust in times of crisis and may to lead to 'break-away' organizations (e.g., cotton producer organizations in Benin). Another issue at stake in the cotton organizations is the fact that the less cotton produced, the lower the amounts collected in levies, which in turn means that real participation declines. This mechanism risks keeping poor farmers poor and making the rich farmers even richer, and thus more involved in decision-making to their advantage.

FUPRO (the national union of agricultural producers in Benin) and MVIWATA (a network of Tanzanian farmer groups) publish journals for their members: in both cases this is a medium for channelling information from the national level down to the grass-roots level. Other mechanisms to inform the membership include the Annual General Meeting (AGM), which all members can attend, and the use of rural radio. All organizations, except ACooBéPA, organize regular AGMs. UCPC Kandi and Djidja contribute to the funding of the district radio station and, in return, use the radio free of charge to inform its members, which proves to be a very efficient means of communication. Strong grass-roots groups that benefit from regular capacity building are important or upward linking, as shown by the experiences of network organizations such as MVIWATA. Grass-roots groups are more eager to have their voices heard if they are also responsible for managing projects such as the *Gacaca* member groups of INGABO. In well-established commodity supply chains that contribute substantially to national economies, member groups of commodity-based organizations (cotton/UCPC) and coffee/KILICAFE) obtain their voicing capacity from the fact that they directly represent the producers. Their position has been reinforced through the withdrawal of state supervision in these chains while 'well-meant' projects (e.g., support for cashew supply by ACooBéPA) may still hamper grass-roots groups from playing their full role.

Social capital

Social capital can be analytically divided into bonding, bridging and linking social capital (Grootaert and Van Bastelaer, 2002; see also Chapter 3, Figure 3). Table 9 provides an overview of the role played by farmers' organizations in bonding, bridging and linking social capital.

In both the network organizations and commodity-based organizations, in terms of bonding and bridging, social capital is very variable since organization forms differ substantially (see also Chapter 5). The basic motivation for farmers to join groups and for groups to join in larger networks and organizations is the fact that collective action is more effective and profitable than individual undertakings. Trust is thereby the main key, since submitting collective action to all kinds of procedures can make it even less efficient. However, more formal procedures are required at higher levels within an organization, whereas local farmer groups often rely on more traditional, village community mechanisms for checks and balances (e.g., the INGABO *Gacaca* groups). Operational transparency and accountability mechanisms within farmers' organizations are therefore crucial. The two Benin cases indicate that badly managed collective action (e.g., weak capacity to negotiate collective marketing contracts for raw cashew nuts) and malfunctioning of governance bodies (e.g., embezzlement of cotton funds) put a strain on social capital. Having the required skills to ensure that the core business of an organization actually functions, is another determining factor for building social capital. Well-defined core functions also make targeting 'partner organizations' easier (e.g., INGABO's relationship with faith-based NGOs that have a clear pro-poor focus).

Some years ago the discussion concerning service providers working with farmer organizations focused on whether researchers or extension workers would be more efficient if working with existing groups (traditional or previously established) or if it would be better if they created 'their own' specific organizations. Today it is widely agreed that more and better results can be achieved if existing organizational forms are taken into account as much as possible (Heemskerk and Wennink, 2004). So far, researchers and extension agents have been quite effective in using bonding to gain social capital, for example for setting research priorities, experiential learning (e.g., through Farmer Field Schools) and disseminating information. However, the potentials available, both in terms of bridging and linking, are under-utilized. In some cases resource persons within groups' communities provide precious social capital for more inclusive service provision to members: e.g., INGABO's farmer facilitators play an important role in farmer-to-farmer extension, and ACooBéPA's initiative to use well-skilled community members to enhance contract negotiations.

Commodity-based farmers' organizations have generally inherited an important potential for linking social capital: they link with both chain actors and public sector agricultural service providers (e.g., KILICAFE and UCPC and their links to extension and research). The main drivers to maintain this linking capital are the performance of the value chain and the quality of the products. With both coffee and cotton, the quality of the final product is taken into account when fixing prices. Network organizations, which focus on lobbying and advocacy, often have to build up their linking capital. They have to 'prove' that they are trustworthy partners, either by representing an important number of members (e.g., MVIWATA) or contributing to services that perform better (e.g., combating the cassava mosaic virus with the help of INGABO members). Network organizations link up mainly with local government, other public sector stakeholders and NGOs.

Farmers' organizations and service provision

The role of farmers' organizations in service provision, and particularly the pro-poor focus of services provided, is strongly determined by both external and internal

Table 9: Role of farmers' organizations in bonding, bridging and linking social capital

Social capital	ACooBéPA Benin	UCPC Benin	INGABO Rwanda	KILICAFE Tanzania	MVIWATA Tanzania
Bonding	Strong village producer groups, exploring and realizing marketing opportunities	Strong village producer groups, registered and crucial for access to inputs and credits	Strong grass-roots institutions, involved in project management	Strong FBGs, registered and essential for central pulping unit	Quality of FGs variable depending on origin Leadership training
Bridging	Relatively weak Lack of adequate facilities for cooperatives	Relatively strong Centrally organized and coordinated cotton chain	Relatively strong Central level, mobilizing resources for village groups	National and Chapters supporting FBGs	Relatively strong at National and intermediate level Inventories for Farmer Fora
Linking	Relatively weak Relations with service providers managed by project and NGOs	Strong, with chain actors and public sector services	Relatively strong at the district level (joint service provision), and central level (through ROPARWA membership)	Strong, with chain stakeholders; less with public sector services	Increasing at national and central levels, but weak locally

Table 10: Factors influencing exclusion/inclusion in service provision for innovation

SWOT analysis	ACooBéPA Benin	UCPC Benin	INGABO Rwanda	KILICAFE Tanzania	MVIWATA Tanzania
Strengths	Small grass-roots membership (trust and affinity) Well-trained membership	Consolidated funding mechanism for all levels Well-trained staff Participation in community development	Grass-roots and intermediate levels participate in service provision Member of national network	Good governance: strong FBGs and leadership Links with research	FGs involved in innovation for market access Lobbying at intermediate and national level
Weaknesses	Weak collective negotiating capacities Little information on markets	Institutionalization of informal eligibility criteria (e.g. cotton production) Lack of transparency in resource allocation	Weak financial sustainability of farmer-to-farmer extension services	Poor gender situation Investment constraints	Poor links at group level Poor demand articulation by groups
Opportunities	Emerging value chain with national policy support Special cashew research programme and entity	Part of a larger established network Links with research and extension at all levels	Links with local authorities and service providers Partnerships with NGOs that have an explicit pro-poor focus	Quality innovation for specialty coffee markets	Farmer Fora at ward and district level for innovation demand
Threats	Agenda setting by donors/projects Lack of appropriate credit facilities	Poor sector governance Declining prices without a clear 'diversification' policy	Donor dependency (financial autonomy) Professionalization of the sector/organization	No pro-poor innovations: labour, (HIV/AIDS) etc. Low external input	FGs weak without demands; efforts do not lead to bridging and empowerment

Table 11: Role of farmers' organizations (FO) and access to research, advisory and training services

Role of FO	ACooBéPA Benin	UCPC Benin	INGABO Rwanda	KILICAFE Tanzania	MVIWATA Tanzania
Accessing services for members	Joint training and extension services nurseries and planting) with projects	Main representative for public sector research and extension services Funding of research and extension (cotton levies)	Farmer-led credit facilities at the grass-roots level Farmer-to-farmer extension at local level	Coffee quality innovation and FBG organization	Lobbying for market projects at national level Lobbying for advice and training at local level
Accessing services for poorest members	Smallholder cashew growers benefit from group training	Smallholder cotton growers benefit from larger farmers (group training)	Selecting farmer extensionists among members Pro-poor projects (donor supported)	Very small coffee producers benefit from larger ones	Increasing role in advisory services for women and HIV/AIDS affected families
Accessing services for non-members	Nurseries are accessible for non-members	Contribution to community infrastructure	Lobbying and advocacy	Lobbying with research for coffee nurseries	Lobbying and advocacy Contracted service provision

factors. Table 10 presents an extract from the different SWOT analyses; these are the key factors that influence service provision (indirectly by third parties) or (directly) by farmers' organizations.

The more internal factors that are related to the organizations themselves include:
1. inclusive eligibility criteria for representing categories of members at levels or platforms where decision-making on service provision is taking place;
2. capacity strengthening and skill development of both the members and staff for adequately voicing of needs and planning services;
3. financial resources and a certain level of financial autonomy as a leverage mechanism for orienting and providing services;
4. building social capital for collective action by members and member organizations, and joint action with service providers.

However, external factors are also important, and these include:
1. national policies for diversifying commodity crops (e.g., niche and speciality markets for cashew and coffee);
2. institutions for voicing farmers' needs (e.g., local farmer fora for priority setting in service delivery);
3. decentralization of agricultural research and extension services;
4. availability of technologies for the poorest.

Farmers' organizations are increasingly taking over services and, as such, provide services to the overall farmer community (both to members and non-members). This tendency is enhanced by overall processes that are going on in the three countries: state withdrawal from providing goods and services, including agricultural extension; decentralization of governance and deconcentration of services; and the lack of funding of public sector services, which forces them to explore new funding mechanisms such as cost-sharing and outsourcing (see Heemskerk and Wennink, 2005). Table 10 presents an overview of the role played by farmers' organizations in service provision. The overall picture shows that farmers' organizations and service providers increasingly work together as a result of public funding constraints and the desire by service providers to gain an 'economy of scale'. However farmers' organizations, particularly network organizations, lobby for 'pulling down' services or start organizing service provision themselves. This offers opportunities for more inclusive and pro-poor services (e.g., 'farmer-led' initiatives, where farmers are involved in all stages of service provision and delivery: targeting services, selecting service workers, and assessing the quality of services).

State withdrawal and decentralization provide farmers' organizations with opportunities but also with challenges. A first challenge is to provide services on a sustainable basis. INGABO's experiences in this field illustrate both the successes (in terms of coverage and reach of different farmer households) and risks (sustainable financing farmer extension services) involved. A second challenge is the need to link up with other knowledge services and sources, because a globalized context and demanding markets require up-to-date information for innovation. Commodity-based organizations, such as KILICAFE and UCPC, therefore still rely on established specialized research organizations. The way way in which demands for services are identified and presented, and knowledge and information are disseminated by the farmers' organization, are crucial. Involving local level groups is therefore essential (e.g., MVIWATA farmer groups) since they have first-hand knowledge of diversity. Although UCPC Benin still very much assumes that knowledge and innovation will eventually diffuse through to all farmers by training the leaders, MVIWATA includes more farmers in training and extension activities.

However, the inclusive character of services provided by research and extension also depends on:
1. the policy context, which may commit service providers to poverty alleviation and a consequent operational strategy;
2. the institutional set up (e.g., level of decentralization of these services);
3. the organizational capacity (e.g., network of agents); and
working methods (e.g., participatory approaches).

We notice that the more private funding that is involved (e.g., through commodity levies or services provided by private enterprises), the more the services are exclusively targeted towards members. The involvement of the public sector in service provision and more community-development-related purposes (e.g., UCPC funding of infrastructure in Benin) seems to be a guarantee for reaching far beyond just the members of a farmers' organization.

6 Concluding remarks: towards a strategy for social inclusion

A need for continuous interaction between agricultural service providers and farmers' organizations

In order to achieve sustainable rural development for the benefit of all categories of rural households, from the poorest to the richer (as well as for all members of these households), the identification of opportunities for viable development is a first requirement. From an innovation perspective, agricultural research and advisory services play an important role in this. Therefore, inclusive access to these services by all referred categories, as well as openness by these services towards the poorest, is therefore central to achieving rapid and sustainable rural development.

The case studies demonstrate the need for a continuous interaction between agricultural service providers and farmers' organizations. This would also allow farmers' organizations to better articulate inclusive demands for which building social capital is an essential condition.

In order to improve access to knowledge services, which is expected to enhance the likelihood that farmers can make use of the opportunities identified, socially inclusive research and advisory services must be available. Different kinds of farmers and categories of households need to be listened to by service providers, in order for their priorities and needs to be included in the development and service agenda, and for the required services to be made widely available as real public goods. This is a continuous and enormous challenge for the research and advisory services systems with which the public, private and 'third' sectors need to interact.

Key roles for farmers' organizations in inclusive innovation systems

Farmers' organizations can play four roles in the pro-poor orientation of services by:
1. lobbying for an enabling policy and institutional environment;

2. facilitating the voice of the poorest and other vulnerable groups to be heard;
3. exercising influence on advancing socially inclusive research and advisory service agendas; and,
4. becoming involved in the implementation of research and advisory services for the poorest and the most vulnerable.

Farmers' organizations can do this on the basis of their mandate for advocacy for the rural poor in general, but also based on the voice of their own constituencies and members. The central questions then are two-fold:
1. do the present types of farmers' organizations have the capacity to strengthen the voice of the poor and actually influence the agenda setting for all categories of households? and
2. do farmers' organizations have the capacity to get involved in service provision, on their own initiative or through contracts with the public (primarily) and private sectors?

The results from both the case studies conducted and the analysis provide us with some strategic elements for capacity strengthening of farmers' organizations for socially inclusive service provision. One of the main conclusions is that farmers' organizations can indeed play strong advocacy and service provider roles, but that a number of conditions need to be met. These conditions mainly relate to capacity development of the farmers organizations at different levels in articulating their needs and demands, and building social capital.

Capacity strengthening of farmers' organizations: articulating inclusive demands

Farmers' organizations, which are primarily involved in production and processing, are central in agricultural innovation. They therefore require capacity development for:
- learning-by-doing and learning-by-interaction. These are key elements in order to strengthen socially inclusive service provision for new technologies and practices.
- enhancing the level of inclusion enhancement for different types of services. However, experiences indicate that this relates to the type of knowledge offered as well as the degree to which the service is considered a public good. The level of cost-sharing of the services provided can also lead to exclusion.
- monitoring of the social inclusiveness of agricultural innovation. In terms (again) of interaction with others, strong performance indicators need to be developed: performance by the actors, their functions and their interaction, as well as with regard to policy-making for socially inclusive and hence sustainable development (see for the agricultural innovation system concept: Wennink and Heemskerk, 2006: pp. 32 and 43-44).

Furthermore, farmers' organizations can undertake specific actions as member-based and member-led organizations:
- Farmers' organizations can develop special programmes to enhance equal opportunities for members to become involved in leadership (at group and higher levels) through skill development and 'learning-by-doing'.
- Farmers' organizations require (and some already have), internal and external policies to advance the interests of women, young members and other vulnerable groups, such as people affected by HIV/AIDS, and specifically on mainstreaming such groups in service provision for innovation development.
- Farmers' organizations can develop their own gender strategies, without leaning towards window-dressing for donors. Gender involves changing cultural values and organizational strategies that help define favourable criteria for access to services and opportunities for women to express their voice.
- Farmers' organizations need to define criteria for the regular elaboration of membership profiles. This will allow the farmers' organization to develop strategies to include special target groups such as young members, households headed by women, HIV/AIDS-affected households, herdsmen, minority ethnic groups etc. More particularly, it allows them to generate innovation-development priorities for each member category and to articulate these accordingly.
- With respect to commodity-based organizations, product quality and related price incentives (instead of bulk quantities) provide an excellent opportunity for poorer farmers to gain a market share and improve their incomes.

Capacity strengthening of farmers' organizations: building social capital

Farmers' organizations that are involved in production and processing are central to agricultural innovation. They therefore require capacity development for the three dimensions of their social capital: bonding, bridging and linking – also in relation to the interaction with all key stakeholders.

Farmers' organizations are most likely to have a socially inclusive membership through strong grass-roots groups. Inclusiveness can be further enhanced through a concentration on more (but not necessarily inclusive) socially mixed groups. Socially mixed groups can exist not only in relation to gender, but also in terms of poverty categories e.g., small and larger farmers in one group, or group member households overcoming stigmas, for example by having households affected by HIV/AIDS included as members. To favour access for the poorest, farmer groups need low thresholds for entry of new members (i.e., limited number and non-exclusive criteria) and active policies to include all types of farmers and rural households.

Bonding social capital is also required to strengthen learning within the community, similar to the approach used in Farmer Field Schools, but with an extra dimension that the poorest and other small-scale farmers, as well as all gender categories, are involved. Farmers' organizations need to develop the internal capacity to strengthen such learning in groups, as well as to exchange experiences between groups, e.g., through farmer motivators, facilitators and farmer group study tours, and between the different tiers of the organization.

Strong bridging of social capital development is essential to achieve closer interactions between the grass-roots level and intermediate/national levels, also in terms of meeting innovation requirements at the grass-roots level and lobbying for an enabling environment at national level. The existing social capital at community level needs to be identified and applied to local networks (i.e., bridging social capital). The strength of the farmer's voice will increase if there are no parallel or competing networks based on social background, gender, ethnicity, or production orientation. On the other hand, networks can be overlapping, as KILICAFE FBGs can also be members of MVIWATA, and ACooBéPA groups are also members of FUPRO.

Similarly, farmers' organizations need to play their role in rural innovation systems, hence the interactive learning role at all levels (local, meso and national); this requires social capital to be linked at these levels. It involves engaging in planning and policy-making, but above all, in serious performance-based monitoring of the research and advisory services being provided. Important elements include the interaction between farmers' organizations and individual group members on the one hand, and for instance extension services on the other, in determining the target villages, groups, individuals and themes.

References

AgroEco, 2006. *Farmer organisation. Status in East and Southern Africa.* Presentation at the second Agri-Profocus Expertmeeting, 22 November 2005. Available online at URL: www.agri-profocus.nl/docs/200703221559085366.pdf and http://agri-profocus.nl/docs/Farmer%20

Barlett, A., 2004. *Entry points for empowerment.* For Care Bangladesh. Available online at URL: www.communityipm.org/docs/Bartlett-EntryPoints-20Jun04.pdf

Beaudoux, E., and M. Nieuwkerk, 1985. *Groupements paysans d'Afrique.* Dossier pour l'action. Paris: L'Harmattan.

Beall, J., and L-H. Piron, 2005. *DFID Social Exclusion Review.* LSE/ODI, London, UK. Available online at URL: www.odi.org.uk/rights/Publications/BBeale&PironSocEx.pdf

Bebbington, A., and J. Thompson, 2004. *Use of civil society organizations to raise the voice of the poor in agricultural policy.* Working Paper. DFID, London, UK. Available online at URL: http://dfid-agriculture-consultation.nri.org/summaries/wp14.pdf

Bingen, J., A. Serrano, and J. Howard, 2003. Linking farmers to markets: different approaches for human capital development. *Food Policy* 28: 405-419.

Bosc, P.M., J. Berthomé, B. Losch, and M.R. Mercoiret, 2002. Le grand saut des organisations de producteurs agricoles africaines: De la protection sous tutelle à la mondialisation. *Recma – Revue des études coopératives, mutualistes et associatives.* No. 285, juillet 2002, pp. 47-62.

Bosc, P.M., D. Eychenne, K. Hussein, B. Losch, M.R. Mercoiret, P. Rondot, and S. Mackintosh-Walker, 2003. *The role of rural producers organizations in the World Bank rural development strategy.* World Bank, Washington, USA.

Chauveau, J.P., 1992. *Le "modèle participatif" de développement est-il "alternatif"?. Eléments pour une anthropologie des "développeurs".* In: Le bulletin de l'APAD, No. 3. Available online at URL: http://apad.revues.org/document380.html

Chilongo, T., 2005. *Tanzanian agricultural co-operatives: an overview.* A draft report. Moshi University College of Co-operative and Business Studies, Moshi, Tanzania.

Available online at URL: www.fredskorpset.no/upload/24657/Tanzanian%20 Agricultural%20Co-operatives%20-%20An%20Overview.pdf

Chirwa, E., A. Dorward, R. Kachule, I. Kumwenda, J. Kydd, N. Poole, C. Poulton, and M. Stockbridge, 2005. *Walking tightropes: supporting farmers' organizations for market access.* Natural Resource Perspectives No. 99, November 2005. ODI, London, UK. Available online at URL: www.odi.org.uk/NRP/99.pdf

Collion, M-H., and P. Rondot, 1998. *Partnerships between agricultural services institutions and producers' organizations: myth or reality?* AgREN Network Paper No. 80. ODI, London, UK.

Debrah, S.K., and E.S. Nederlof, 2002. *Empowering farmers for effective participation in decision making (Benin, Burkina Faso, Ghana and Mali).* Report IFDC-Africa. IFDC, Lomé, Togo.

DFID, 2004. *What is pro-poor growth and why do we need to know?* Pro-poor Growth Briefing Note 1, DFID. Available online at URL: www.dfid.gov.uk/pubs/files/propoorbriefnote1.pdf

DFID, 2005. *Growth and poverty reduction: the role of agriculture.* A DFID Policy Paper. DFID, London, UK. Available online at URL: www.dfid.gov.uk/pubs/files/growth-poverty-agriculture.pdf

DFID/FAO, 2000. *Inter-agency Experiences and Lessons.* From the forum on operationalizing Sustainable Livelihoods Approaches. Pontignano, Siena, 7-11 March 2002. DFID/FAO, London/Rome, UK/Italy.

Diagne, D., and D. Pesche (eds.), 1995. *Peasant and rural organizations: forces for development in Sub-Saharan Africa.* French Ministry of Cooperation, Paris, France.

Docking, T.W., 2005. International influence on civil society in Mali: the case of the Cotton Farmers' Union, SYCOV. Chapter 8 in: Igoe, J., and T. Kelsal. 2005. *Between a hard rock and a hard place.* African NGO's, donors and the state, pp. 197-222. Carolina Academic Press, Durham, USA.

Eames, M. with M. Adebowale (eds.), 2002. *Sustainable development and social inclusion: towards an integrated approach to research.* Policy Studies Institute, Joseph Rowntree Foundation, York. Available online at URL: www.capacity.org.uk/assets/downloads/JRFSocial_exclusion_and_SD.pdf

FAO, 2007. *Gender and food security.* Available online at URL: www.fao.org/Gender/en/agrib4-e.htm

Farrington, F., 2002. Towards a useful definition: advantages and criticism of 'social exclusion'. In: *The journal of Geos: Geography, Environment, Oekumene Society (GeoView),* Flinders University, Australia. Available online at URL: www.ssn.flinders.edu.au/geog/geos/farrington.html

Gallagher, K.D., 2001. *Semi-self financed field schools and self financed field schools: helping farmers go back to school in IPM/IPPM.* Unpublished report, prepared by Kevin D. Gallagher based on the work of an IFAD support programme. Available online at URL: www.farmerfieldschool.net/document_en/28_29.pdf

Goetz, A.M., and J. Gaventa, 2001. *Bringing citizen voice and client focus into service delivery*. IDS Working Paper 138. IDS, Brighton, UK.

Grootaert, C., and T. Van Bastelaer, 2002. *Understanding and measuring social capital: a multi-disciplinary tool for practitioners*. World Bank, Washington, USA.

Gubbels, P., and C. Koss, 2000. *From the roots up. Strengthening organizational capacity through guided self-assessment*. World Neighbors Field Guide 2. World Neighbors, Oklahoma City, USA.

Heemskerk, W., and B. Wennink, 2005. *Building social capital for agricultural innovation: experiences with farmer groups in Sub-Saharan Africa*. Bulletin 368. Development Policy and Practice, Royal Tropical Institute (KIT), Amsterdam, The Netherlands. Available online at URL: www.kit.nl/smartsite.shtml?id= SINGLEPUBLICATION&ItemID=1760

IDS, 2006. *Livelihoods Connects. What are livelihoods?* Available online at URL: www.livelihoods.org/SLdefn.html

IFAD, 2007. *Rural poverty in Africa*. Available online at URL: www.ruralpovertyportal.org/english/regions/africa/index.htm

Irz, X., L. Lin, C. Thirtle, and S. Wiggins, 2001. Agricultural productivity growth and poverty alleviation. *Development Policy Review* 19 (4): pp. 449-466.

Jackson, A., 2001. Social inclusion/exclusion of Canadian children: In: *Horizons* (4:1). Ottawa. Projet de recherche sur les politiques, février, pp. 4-5.

Kabeer, N., 2001. Discussing women's empowerment. Theory and practice. *SIDA Studies* No. 3, pp. 17-54.

MacDonald, M., E. Sprenger, and I. Dubel, 1997. *Gender and organizational change. Bridging the gap between policy and practice*. KIT Publishers, Amsterdam, The Netherlands.

Messer, N., and P. Townsley, 2003. *Local institutions and livelihoods. Guidelines for analysis*. Rural Development Division, FAO, Rome, Italy. Available online at URL: www.fao.org/docrep/006/y5084e/y5084e00.htm

Mukhopadhyay, M., and M. Sultan. 2005. *Human rights and good governance review of Danida Sector Programme (Bangla Desh)*. Development Policy and Practice, Royal Tropical Institute (KIT), Amsterdam, The Netherlands.

Nederlof, E.S., 2006. *Research on agricultural research. Towards a pathway for client-oriented research in West Africa*. Published doctoral dissertation. Wageningen University, Wageningen, The Netherlands.

North, D.C., 1990. *Institutions, institutional change and economic performance*. Cambridge University Press, Cambridge, UK.

OECD, 2006. *Promoting pro-poor growth. Agriculture*. DAC Guidelines and Reference Series. A DAC Reference Document. OECD, Paris, France. Available online at URL www.oecd.org/dataoecd/9/60/37922155.pdf

Øyen, E., 2001. Six questions to the World Bank on the World Development Report 2000/2001: "Attacking poverty". Chapter 2 in: *A critical review of the World Bank report: World Development Report 2000/2001. Attacking poverty*. Comparative research programme on poverty. Available online at URL: www.crop.org/publications/files/report/Comments_to_WDR2001_2002_ny.pdf

Pesche, D., 2001. *Classification et typologies des organisations paysannes*, AGRIDOC, Inter-Réseaux Développement Rural, Paris, France.

Roesch, M., 2004. *BIM: L'analyse des organisations paysannes?* For Espace Finance, France. Available online at URL: http://microfinancement.cirad.fr/fr/news/bim/BIM-2004/BIM-21-04-04.pdf

Sen, Amartya K., 1981. *Poverty and famines.* Clarendon Press Oxford, UK.

Shirbekk, G., and A. St. Clair, 2001. A philosophical analysis of the World Bank's conception of poverty. Chapter 3 in: *A critical review of the World Bank report: World Development Report 2000/2001. Attacking poverty.* Comparative research programme on poverty. Available online at URL: www.crop.org/publications/files/report/Comments_to_WDR2001_2002_ny.pdf

Shookner, M., 2002. *An inclusion lens. Workbook for looking at social and economic exclusion and inclusion.* Population Health Research Unit, Dalhousie University. Social Inclusion Reference Group, Atlantic Region. Available online at URL: www.phac-aspc.gc.ca/canada/regions/atlantic/Publications/Inclusion_lens/inclusion_2002_e.pdf

Shrestha, Nanda R., 1990. *Landlessness and migration in Nepal.* Boulder: Westview Press USA.

Silver, D., 2004. *Reaching the poor: bring services closer to the poor.* Mellemfolkeligt Samvirke. Newsletter 3/2004 June. On reaching the poor. Available online at URL: www.ms.dk/sw9616.asp?usepf=true

Toye, P.M., and J. Infanti, 2004. *L'inclusion sociale et le développement économique communautaire: Recension des écrits.* Réseau canadien de DEC. Available online at URL: www.ccednet-rcdec.ca/fr/docs/rpadc/PCCDLN_20040803_RecensionB.pdf

Wennink, B., and W. Heemskerk, 2006. *Farmers' organizations and agricultural innovation. Case studies from Benin, Rwanda and Tanzania.* Bulletin 374. Development Policy and Practice, Royal Tropical Institute (KIT), Amsterdam, The Netherlands. Available online at URL: www.kit.nl/smartsite.shtml?ch=fab&id=SINGLEPUBLICATION&ItemID=1965

World Bank, 2001. *World Development Report 2000/2001. Attacking poverty.* Available online at URL: http://siteresources.worldbank.org/INTPOVERTY/Resources/WDR/overview.pdf

World Bank, 2006. *Enhancing agricultural innovation: how to go beyond strengthening of research systems.* Agricultural and Rural Development, World Bank, Washington, US. Available online at URL: http://siteresources.worldbank.org/INTARD/Resources/Enhancing_Ag_Innovation.pdf

World Bank, 2007. *World Development Report 2008: Agriculture for development.* Available online at URL: http://econ.worldbank.org/WBSITE/EXTERNAL/EXTDEC/EXTRESEARCH/EXTWDRS/EXTWDR2008/0,,menuPK:2795178~pagePK:64167702~piPK:64167676~theSitePK:2795143,00.html

Part II
Case studies on the role of farmers' organizations in accessing services

1 INGABO's role in pro-poor service provision in Rwanda

Jean Damascène Nyamwasa and Bertus Wennink

Introduction

Rwanda is a poverty-stricken country: 60% of the people live below the poverty line and 40% live in extreme poverty. Poverty is above all a rural phenomenon, with more than 90% of the poor living in rural areas. Due to the 1994 genocide, in 2001 around 30% of the households in Rwanda were headed by women (this was only about 20% in 1990). HIV/AIDS infection is particularly prevalent among pregnant women. The overall HIV/AIDS prevalence is approximately 5% of the population; but it is twice as high in urban areas than in rural areas (IFAD/MINAGRI, 2004a and 2004b; OECD, 2006). Poverty is strongly related to ownership of land and cattle. The poverty alleviation strategy in Rwanda distinguishes six categories of households according to the level of poverty (see Table 1).

Table 1: Categories of households according to their poverty level

Category	Food self-sufficiency	Land	Cattle	Source of income	Savings
'Destitute'	No	No	No	Begging	No
'Very poor'	No	No	No	Incidental labour jobs	No
'Poor and self-sufficient'	Yes	Yes	Some	-	No
'Poor and able to save money'	Yes	Yes	Yes	Marketing surplus production	Yes
'Rich and self-sufficient'	Yes	Yes	Yes	Marketing surplus production Sometimes salaried jobs	Yes
'Rich and able to save money'	Yes	Yes	Yes	Marketing surplus production Salaried jobs	Yes

Source: Government of Rwanda (2002: p. 15) and IFAD/MINAGRI (2004a).

The 'Destitute' are estimated as 11.5% of households, while households with less than 0.2 ha of land are estimated at 28.9% (Government of Rwanda, 2002: p. 17).

Rwanda has a long-standing tradition of farmers organizing themselves into groups. Traditional mutual-aid groups for farm work and building houses have always existed and still operate. Farmers also join together to handle input supply, store and market products, manage savings and credit schemes, develop inland valleys, etc. A large number and variety of farmers' organizations currently exist: associations and cooperative-type groups *(groupements de base)* at the local level; and loosely organized networks of several associations and/or cooperatives *(intergroupements)*, cooperative unions, producer federations *(fédérations)* and farmer unions *(syndicats)* at higher levels. Private and state-owned enterprises have also created cooperative-type organizations to organize the marketing of commodities such as coffee or tea (Bingen and Munyankusi, 2002; MINAGRI/CTB, 2005).

In fact, two major factors explain the rapid expansion of farmers' organizations in Rwanda. First there were the incentives from state services and related development projects to form farmer groups. Such incentives include promising access to inputs, credit facilities and inland valley irrigation schemes. The second factor (at a later time) was the state's withdrawal from service provision and the privatization of public sector activities. This obliged farmers to organize themselves on their own initiative, in order to facilitate access to services, facilities and marketing of products. The pressure on farmers to organize themselves into groups was further enhanced by the devastating effects of the genocide at the beginning of the 1990s, which left the infrastructure (e.g., for seed and input supply) severely damaged, and left basic agricultural services (e.g., extension services) non-operational. Once the genocide ended in 1994, many associations and cooperatives were reorganized, or newly formed, on the initiative of the farmers themselves and/or with help from NGOs, as a means of stimulating self-help among farmers and their communities. During this same period, new types of farmers' organizations, such as INGABO, emerged at provincial and national levels; often with support from NGOs and donor agencies.

At present, associations and cooperative-type groups are widely spread. Such associations and groups often combine economic and social objectives, and cover a wide range of activities. They are registered with the district authorities under the law on cooperatives. Several associations and groups may join to become a cooperative, or even a cooperative union and, as such, are being recognized by the ministry responsible for cooperatives. However, as yet, few cooperatives have been registered and recognized by the ministry. This is probably because most cooperatives lack the resources and managerial capacities to advance rapidly (Bingen and Munyankusi, 2002). A survey among farmers' groups at the grass-roots level showed that at least 50% of all members of all associations and farmers' groups are female, and that more than 20% of the members are literate (Ibid: pp. 2-3). This same survey indicates that public sector services rely on both the farmers' organizations and NGOs to implement government policies. In this respect, the government policy for the agricultural sector further acknowledges the role of farmers' organizations (MINAGRI, 2004).[1]

Farmers' unions and producer federations aim to defend the interests of farmers at levels beyond those planned by *groupements* and *intergroupements*. Producer federations in Rwanda are often linked to specific supply and value-chains, while farmers' unions are oriented towards more general issues. The best known federations and unions in Rwanda are IMBARAGA, a federation of farmers and livestock holders, FERWATHE, a federation of tea growers, and the *Syndicat Rwandais des Agriculteurs et des Eleveurs* INGABO, a union of farmers and livestock holders. Both INGABO and IMBARAGA are members of ROPARWA, a national network that regroups several federations and unions.

This case study deals with INGABO that, besides IMBARAGA, is one of oldest farmer unions in Rwanda, with considerable membership. In 2006, interviews were conducted with both members and leaders; field visits were also made to explore the ways in which INGABO addresses the issue of social inclusion for service provision. Results were discussed during a small workshop with members and leaders.

Presenting INGABO

INGABO, which means 'army' in the Kinyarwanda language[2], is a farmers' union that was created on January 17, 1992, by farmers and livestock holders from Gitamara province, one of the former 12 provinces in Rwanda.[3] Its establishment took place within a particular context of political tensions (a prelude to the genocide): farmers felt threatened in their social cohesion by politicization and political alignments. The founding farmers felt they needed an organization that centred on their interests. Since then INGABO has evolved towards becoming a legally recognized organization, a union; for details see Table 2.

INGABO's mission is to defend the economic interests of its members, both farmers and livestock holders by:
1. uniting them to become a force that is capable of defending their interests;
2. training members to improve their professional activities;
3. defending the honour, justice and unity of farmers and livestock holders, as well the cooperation among them;
4. participating in preparing decision-making on issues that concern farmers and livestock holders, and in such decision-making itself (notably in relation to land tenure);

Table 2: INGABO's trajectory towards becoming an officially recognized farmer union

Period	Key events and developments
1992-1993	Founding of INGABO Growth in membership Functioning on a voluntary basis
1994	Activities halted due to the genocide Members fled or were killed Destruction of infrastructure
1995-1997	Reconstruction of the organization Support from, and intensive collaboration with, local NGOs INGABO is increasingly perceived as a support organization for farmers In some way INGABO becomes a hybrid organization with NGO features
1998-2004	Institutional development and organizational strengthening with NGO support Definition of strategic orientations Setting up governing bodies and management structures Recruitment of salaried staff
2005 onwards	Revision of policy and strategic orientations as a farmers' union (*syndicat*) Obtaining a legal status with the Ministry of Labour

5. encouraging and supporting members to contribute to social and community development; and,
6. promoting agriculture for national development.

The basic entity of INGABO at the local level is the '*Gacaca*'[4], which organizes member meetings and consists of 20 members but does not have a legal status. It is governed by an elected committee *(bureau local)*, which in turn is controlled by a supervising committee *(collège de surveillance)*. This same organizational model is repeated at the district and province levels within INGABO. The local level also has a mediation committee *(comité de médiation)*, which mediates and arbitrates in cases of disagreements between members. At the central level, INGABO has a general assembly *(congrès général)*, board of directors *(conseil supérieur)*, committee of directors *(bureau directeur)*, and a technical committee *(coordination technique des activités)*. The union has technical staff at the central level, including agents for training and advising members. They are represented in the technical committee.

In 2006, INGABO counted 12,983 members: 7,430 of these are women. Members are organized in *Gacacas*: about 800 *Gacaca*s that cover the 1,097 cells *(cellules*[5]*)* of the former Gitamara province (covering over 70%). Members pay membership fees and are individual INGABO 'cardholders' who participate in meetings and contribute to union activities.

In March 2006, the general assembly decided to explore new ways to improve the economic position of members and reinforce the resource base of the union by:
1. enhancing the profitability of farming through improving the use of land and labour for farms as business entities;
2. improving profits through networking with other enterprises for better marketing and sale of products;
3. increasing investments in agriculture and livestock holding through mobilizing resources; and,
4. expanding the membership base by involving members and encouraging other farmers and livestock holders to join.

INGABO provides several services to its members; the main ones being:
- Lobbying, advocacy and defending the interests of members when dealing with government authorities, the public and private sector, and civil society organizations.
- Facilitating access for its members to credit and savings facilities (local banks, mutual financing institutions).
- Organizing producers around agricultural supply- and value-chains.
- Training members in the use of both agricultural technologies and farm management techniques.
- Formulating development projects for members and mobilizing resources for their implementation.

Membership of INGABO

The founding statutes of INGABO state that every farmer and livestock holder who signed the initial establishment texts of the organization, as defined by the INGABO assembly in March 1996, is a founding member *(membre fondateur)*. This same text states that every member farmer or livestock holder who agrees with the union's statutes and who has received an endorsement from the governing bodies of the union can become an affiliated member of INGABO. According to INGABO's rules, basic prerequisites for becoming a member of INGABO are: being a farmer or livestock holder (whether land is owned or rented) and (mostly) full-time active in agriculture or livestock holding and earning at least 75% of one's revenues from agriculture. The procedure for becoming a member includes writing an application letter to the local *Gacaca*, which provisionally accepts such applications while waiting for final acceptance by the general assembly and paying a membership fee of 1,000 Rwandan francs.[6] Besides the annual membership fee, every member also has to pay a monthly contribution of 400-450 Rwandan francs to a *Gacaca* fund. This is a savings fund *(tontine)* that is used to give small credit amounts to *Gacaca* members or to support members who face difficulties.

All members of INGABO have the right to participate in the official INGABO meetings, elect their representatives in governing bodies, and are eligible to apply for positions in these bodies. According to the organization's statutes and rules, the union assists members in protecting their professional rights and defending their interests. They are being defended when exercising their profession or undertaking INGABO activities. On the other hand, members have certain obligations towards INGABO: financial contributions, participation in meetings and involvement in other INGABO-sponsored activities. Not respecting the union's rules may lead to a temporary membership suspension or even cancellation.

Gender dimensions

Women form up to 60% of the INGABO membership and, as a result, they account for the majority of members in most of the *Gacacas*. As previously explained, the number of female-headed households has risen since the 1994 genocide and, like their male counterparts, women farmers also seek to have access to production factors to improve their livelihood conditions. Women (widows) have the same heritage rights as men as long as their marriage is officially recognized. Rules and procedures adapted by INGABO encourage women to become members of the union. The statutes and by-laws of the association promote participation by women in governing bodies and surveillance committees.

These institutionalized measures are the result of a two-stage process within INGABO. During the first stage, specific women-related activities were undertaken to take into account the position and concerns of rural women. For example, the establishment of specific union bodies to represent women members' interests, plus the sensitization and training of women on property rights and on access to small cash credit facilities for income-generating activities. This allowed women to improve their socioeconomic situation, adopt a higher profile through income generation, and raise their voices within INGABO. Women-specific representative bodies and activity programmes helped to put women's concerns on the agenda, but also tended to isolate them somewhat within the union.

As a reaction to this, and in a second stage, INGABO's women-specific bodies were dismantled to make way for more inclusive and incentive-oriented measures. These measures aimed to mainstream gender issues within the INGABO governing bodies, in services provided and in activities undertaken by INGABO. For example, the union's by-laws now prescribe a female quota for elected bodies (50% of the seats in the governing bodies should be held by women), and in monitoring and evaluation of services provided (these include 'output indicators' concerning the participation of women). Monitoring and evaluation data currently indicate that women's participation levels to be about 40% in decision-making and development activities.

Building social capital

The INGABO founders consider their union to be an organization that offers opportunities to their members to 'value' themselves as members of a professional corps and to gain societal esteem. Over the years, members have benefited from capacity building and skill development activities in areas such as leadership, organizing and presiding over meetings, communication, social counselling, and conflict resolution. These are skills that are highly valued, for example, during elections for community-based organizations. In fact, many INGABO leaders are also (opinion) leaders at the local level. In 2005, all members of INGABO's general assembly (75 members) held elected positions in local governance bodies at the cell and sector levels (e.g., committees for community development, conciliation committees, and local genocide courts[7]) because their skills were acknowledged.

Collaboration between INGABO and partners is also taking place at other levels: *Gacacas* and other union entities discuss, negotiate and cooperate with authorities and service providers, such as the agricultural extension service (the provincial *Direction Agriculture, Elevage et Forêts* and the district *Responsable du Service Agricole du District*) and agricultural research (the *Institut des Sciences Agronomiques du Rwanda*), on behalf of their members. INGABO has found that communication with authorities and services becomes more straightforward when trust has been built up through successful joint activities (see Box 1).

Box 1: INGABO's role in combating the cassava mosaic virus

As part of a support programme for value chain development, INGABO organized cassava growers into a federation that facilitates and coordinates supplies of cassava cuttings and marketing of cassava. It has about 2,000 members who each grow at least 0.2 ha of cassava. After a devastating epidemic of cassava mosaic virus, all breeders of manioc cuttings who were members of INGABO agreed to provide fields, use healthy cuttings supplied by agricultural research (ISAR), and to be trained in appropriate technologies. Their effective application of these measurements and the consequent impact made cassava growers achieve recognition from both local authorities and agricultural services. INGABO leaders at the different levels of the organization also participated in the evaluation of the programme.

Schrader *et al.*, 2006.

INGABO's contribution is considered particularly successful in three activities: (i) providing credit through the Gacaca funds; (ii) organizing farmer-to-farmer extension; and, (iii) creating producer federations around crops such as fruits and cassava. The producer federations receive technical support from INGABO to develop into autonomous cooperatives, but all members are also INGABO member cardholders.

At the central level, INGABO, as a farmer union and through its membership of the ROPARWA network of farmers' organizations, is also a member of international farmers' organizations (e.g., the East African Farmers Federation and the International Federation of Agricultural producers), and receives support from NGOs (e.g. The Agriterra Foundation in The Netherlands) and donor agencies. INGABO is also solicited by the Rwandan Government to participate in the implementation of the national agricultural policy (MINAGRI, 2004).

Representativeness of INGABO

According to the INGABO's official statutes, only smallholder farmers can become members (i.e., the majority of farmers in Rwanda). An estimated 60% of cell members are also members of a *Gacaca* in areas of the Gitamara province where there is a high concentration of INGABO members. A survey of member records indicates that the 'destitute' and the 'very poor' who don't own land (see Table 3), are not represented among the INGABO membership.

Table 3: Membership of INGABO according to farm holding

Area of farm holding	Percentage of INGABO members
More than 1.5 ha	20%
Between 1 and 1.5 ha	40%
Between 0.5 and 1 ha	25%
Less than 0.5 ha	15%

Both the membership criteria and financial contributions to which membership is subjected, exclude rural people who farm (owned or rented) land and/or make their living from hiring themselves out as farm workers. These are people who don't earn at least 75% of their income from actual farming, and usually cannot pay the required monthly contribution to the *Gacaca* fund. Consequently the question is whether INGABO still takes the concerns of these neglected social categories into account. On the other side of the social spectrum, civil servants and traders, who often earn more than 75% of their income from agriculture, are also being excluded. However, they do not spend most of their time on agricultural activities (farming or livestock holding). INGABO assumes that these groups are insufficiently committed to the union's causes. The union also fears that membership by civil servants and traders may lead to a 'conflict of interest', since these social groups are active in economic sectors other than agriculture.

Roles of farmers in INGABO

A quick scan of the composition of the committee of directors *(bureau directeur)* shows that larger farm holdings are better represented (see Table 4).

Table 4: Composition of elected bodies according to farm holdings[8]

Area of farm holding	Percentage of committee members
More than 1.5 ha	40%
Between 1 and 1.5 ha	50%
Between 0.5 and 1 ha	10%
Less than 0.5 ha	0%

To become elected as a member of one of the governing bodies at the central level, a candidate needs be a member of the general assembly. To become an elected representative in the general assembly, a candidate needs to be a member of a *Gacaca* governing committee *(bureau local)*. Once a person has been elected representative in one of the central governance bodies, he/she must resign from the position in elected bodies at lower levels.

In-depth interviews with INGABO leaders and members provided some additional background information. In general, holders of larger farms show little interest in the activities and services provided by INGABO. They have access to their own resources and services and often feel socially superior to smallholders. But those holders of larger farms who do decide to become union members often become interested in governing the organization and are able to get themselves elected as representatives because they are considered 'role models' of successful members of society. Smallholders often lack self-confidence and consequently underestimate their leadership capacities. They still struggle to fulfil their basic needs such as food, lodging and healthcare. Even if they stand as a candidate for elected positions, their peers generally do not see them as model farmers.

Most INGABO leaders come from the intermediate category of farmers because they are trustworthy and 'recognizable' representatives for the majority of INGABO members. The main criteria considered when it comes to electing representatives include: overall behaviour within the community, being sensitive to social aspects, communication skills, and the quality (and promotion) of novel ideas during meetings. Being a model agriculturalist (applying appropriate new technologies – the main criterion for being a model agriculturalist) is also a criterion for leadership, but is less important. Leadership skills are clearly considered more important than farm management and innovation capacities. However, the latter skills are very important when it comes to selecting farmer extension workers among union members.

SWOT analysis

Table 5 presents an overview of the strengths and weaknesses of INGABO, as well as the opportunities and threats that its environment provides, according to members and leaders who have been interviewed.

The risks of social inclusion are twofold:
- At the national policy level, a programme to make the agricultural sector more professional, brings to the fore those rural entrepreneurs who present guarantees for investments, such as owning land. This excludes the landless or those who make their living on borrowed land, and transforms them into a class of rural labourers. However, doubts exist about viable sectors or activities that may absorb this growing social category in the near future. Neither national policies nor INGABO strategies foresee any real alternatives for them.
- As for INGABO itself, smallholders are relatively weakly represented within the elected governing bodies. This observed tendency seems to continue and may lead to further exclusion of this social category. However, smallholders form up to 40% of the INGABO membership. Their voice may not be sufficiently heard when defining and implementing strategies for agricultural development. If they are not physically represented in governing bodies at higher levels within the union, the challenge is to make their voices heard and their concerns taken into account.

Role of INGABO in access to services

INGABO particularly strives to better organize its information, training and advisory services for agricultural development, and to make farming more professional. These services aim to improve both livestock holding and crop farming, notably for cash crops. Organizing these services was primarily a strategy to fill the gaps left by declining public sector agricultural services. Since 2002, these services are being provided through a network of voluntary farmer extension workers who are either specialized in crop farming *('karemano')* or livestock breeding *('mwungeri')*. Almost all *Gacacas* have elected two farmer extension workers among their members (see Box 2). Currently the INGABO network currently has some 1,200 of these voluntary agents. These volunteers train other *Gacaca* members and receive support from two INGABO-paid agronomists and a livestock specialist, who in turn collaborate with agricultural research and extension services.

Table 5: INGABO SWOT analysis for social inclusion

Strengths	Weaknesses
INGABO is anchored at the grass-roots level through *Gacacas*; members are acquainted with the Union.	INGABO has little financial resources of its own and is still largely dependent on external funding.
Gacacas are small and embedded in communities; members know each other and dare to speak out.	Members feel service provision is still deficient; there is a lack of adequate M&E systems.
Structures and procedures are flexible; representation and participation of all is therefore possible.	Weak representation of smallholders in elected governing bodies; an ongoing process of social exclusion.
Bottom-up strategic and operational planning; taking in account both needs and novel ideas that exist on the ground.	No specific programmes for HIV/AIDS-affected members, which are an important category to be considered.
Gender mainstreaming is seen as successful; women are represented in decision-making bodies.	
Local leaders are involved in selecting beneficiaries for pro-poor programmes.	
Opportunities	**Threats**
INGABO has a track record of activities and is acknowledged as a local development actor.	INGABO still largely depends on external funding; donors may also set the agenda.
The Rwandan Government's decentralization policy, which delegates decision-making to local levels.	Ongoing professionalization of the organization may limit grass-roots participation and voicing.
Growing number of rural radio stations; a good means of communication and information.	Ongoing professionalization of the agricultural sector; a growing category of landless people who are hardly ever represented.
Support for organizational strengthening of the organization; improving (external) accountability.	
Presence of externally funded pro-poor programmes with a clear focus on the poorest.	

INGABO provides two more services that are considered to have a direct impact on farmers' livelihoods. First, there is the artificial insemination of cows in collaboration with the National centre for artificial insemination (CNIA). Over 900 cows were inseminated in 2005. However, until now, only one-third of the inseminations has been successful. Cows play an important role in farmer households by providing manure and milk, and contributing to improving farmer livelihoods. Secondly (and an undeniable success) is the network of local mutual savings and credit schemes, at the *Gacacas* level. Each *Gacaca* member contributes 400-450 Rwandan francs per month to a 'savings fund' and 40-50 francs to a 'mutual fund'. The savings fund provides credit to members for all kinds of activities and investments, while the mutual fund assists members when they face social or monetary problems (see Box 3).

The *Gacaca* fund is an example of INGABO's support to advancing the combination of social and economic development objectives. A major constraint in developing these funds further is the limited financial capacity of their members to provide contributions. A more formal network of savings and credit schemes *(Caisses Locales d'Epargne et de Crédits Agricoles Mutuels)* has been put in place to allow the *Gacaca* funds to be replenished. These micro-finance institutions are managed according to the regulations of the Rwandan national bank. However, formal procedures for mobilizing and accessing funds through the system of agencies may disturb the local social dynamics and solidarity, as well as the pragmatic and tailor-made approach that characterizes the *Gacacas*.

> **Box 2: Selecting and evaluating INGABO farmer extension workers**
>
> Farmer extension workers are selected from *Gacaca* members who are: (i) well-established and innovative farmers; (ii) good communicators; and, (iii) available to give advice to other farmers. They are not paid for their services but receive an allowance when they follow training sessions. Other incentives are training, exchange visits and social esteem. Another incentive is a competition among farmer extension workers who are being evaluated, according to a defined evaluation grid, by union members at all levels. Winners receive prizes (in kind) such as farm equipment and fertilizer. A main challenge for INGABO is to maintain the knowledge level of such a dense network and keep it financially sustainable.
>
> Schrader *et al.*, 2006.

The linking with formal banks is an example of complementary action taken by the central level of INGABO. The *Gacacas* are the grass roots and basic INGABO platform that plans the activities and services to be provided to members. However, a *Gacaca* may plan beyond its members' resources and means, and that is where INGABO (as a national union) comes in. Activities and services that require extra resources are submitted for endorsement to the central committee, where they are integrated into central planning. The central level, through its membership of national (e.g., ROPARWA) and international networks and contacts with NGOs and donor agencies, is well situated to mobilize additional funds. Experience indicates that clarity of plans and competencies at the intermediate level are determinant for effective communication between the grass-roots and central levels.

Box 3: The *Gacaca* fund of Nyagasozi (Sector of Rukoma, District of Kamonyi)

Members	21 women + 16 men
	Among them: 6 widows + 5 orphans
	All members have at least one goat and most of them also have one cow
	Elected committee: 5 women + 4 men
	Supervising committee: 2 women + 1 man
	Mediation committee: 2 men + 1 woman
	Management committee of the fund: 2 women + 2 men
	Fund control committee: 3 men
Activities	Vegetable farming in an inland valley
	Land (53 hectare) is rented from district authorities
	Each member has his/her own field
	Collective procurement of inputs and individual marketing of vegetables
Member contributions to the fund	400 Rwandan francs/month to the credit fund
	40 Rwandan francs/month to the mutual fund
	600 Rwandan francs/year to rent the field
Types of credits provided	Membership fee for the local health mutual
	Payment of school fees for children
	Funding agriculture and livestock activities
	Investing in land or animals (procurement)
	Repair and construction of houses
Types of subventions given (mutual)	Funeral of close relatives
	Marriage
	Daily allowances for leaders when on mission (500 Rwandan francs/person/day)
INGABO contribution	Three irrigation pumps and vegetable farming equipment and two cows
Current balance of the fundLie	550,000 Rwandan francs
	124,000 Rwandan francs through interest on credits
Fund replenishment	800,000 Rwandan francs by CLCAM that has been reimbursed
Beneficiaries of the fund	Women and orphans are given priority

Bottom-up planning of activities and service provision also draws INGABO's attention to formulating plans and mobilizing resources for programmes that provide services to the poorest of their members. Poverty is a common condition in rural Rwanda and most *Gacaca* members face the conditions and consequences of being poor on a daily basis. A special programme, supported by a faith-based NGO, aims to support female-headed households (widows) by providing them with small livestock (goats or pigs), and vegetable seeds. In return for programme support,

INGABO provides training and demonstrations through its farmer extension workers. The *Gacaca* is entirely responsible for identifying the most vulnerable women who participate (Church World Service, 2005).

Concluding remarks

INGABO's experience shows that small, community-embedded grass-roots farmer groups, the *Gacacas*, can be an important factor in enhancing farmer participation in both planning and implementation. Farmer participation and voicing is further increased through INGABO's explicit policy to develop members' capacities and by making them take part in service provision to members.

Participation by all members, particularly the poor, is fostered through integrating social objectives into the organization's service provision. The *Gacaca* mobilize savings funds that have both economic and social functions. In fact, this is a way to reduce the effects of vulnerability among certain social groups. For the poorer farmers, membership of a *Gacaca* is a way of subscribing to an insurance policy. This also explains the success of these funds and requires a careful approach when formalizing grass-roots financial institutions.

The inclusiveness of INGABO and the services provided are enhanced by devolving responsibilities for activities and services to local levels (e.g., selection of beneficiaries). INGABO's central level reinforces the existing mechanisms by mobilizing supplementary external resources for development activities. These activities prove to be successful and pro-poor, because selecting beneficiaries is left to the local level, which has its own selection criteria that are accepted by the local communities.

Exclusiveness of the organization is primarily a result of membership criteria such as being a farmer or livestock holder, and thus having land, and earning a substantial part of one's income from agriculture. Not all rural farmers qualify on the basis of these criteria, but INGABO still insists on applying these criteria. The dominating political discourse of 'professionalization' (both outside and inside the union) could further block membership of INGABO. Secondly, smallholder members are currently excluded because they have difficulties in getting elected to governing bodies – they do not have the time and resources for campaigning or governing, and they are not considered social role models by their peers.

The success in mainstreaming gender may provide an approach for mitigating exclusion mechanisms. Mandated committees and specific procedures could be put in place to meet the needs faced by the poorest, define targeted activities to be undertaken, and pave the way for mainstreaming inclusiveness within INGABO. Furthermore, opening up membership to the growing number of rural labourers within the production value chains to be developed, could provide a response to their needs.

References

Bingen, J., and L. Munyankusi, 2002. *Farmer associations, decentralization and development in Rwanda: challenges ahead.* Agricultural Policy Synthesis. Rwanda Food Security Research Project MINAGRI. No. 3E, April 2002. Available online at URL: www.aec.msu.edu/fs2/rwanda/PS_3E.PDF

Church World Service, 2005. *Hunger & Development. Malnutrition & Poverty Reduction, Rwanda.* Available online at URL: www.churchworldservice.org/Development/project_description/descriptions/91.html

Government of Rwanda, 2002. *Poverty reduction strategy paper.* Available online at URL: www.imf.org/external/np/prsp/2002/rwa/01/063102.pdf

IFAD/MINAGRI, 2004a. *Agriculture in Rwanda. Background.* Bibliographical document compiled within the framework of the activities of IFAD in collaboration with MINAGRI. Text and design by C. Bidault. Available online at URL: www.ruralpovertyportal.org/english/regions/africa/rwa/agriculture_in_rwanda/CDRom/present/english/backgr.ppt

IFAD/MINAGRI, 2004b. *Agriculture in Rwanda. Current situation.* Bibliographical document compiled within the framework of the activities of IFAD in collaboration with MINAGRI. Text and design by C. Bidault. Available online at URL: www.ruralpovertyportal.org/english/regions/africa/rwa/agriculture_in_rwanda/CDRom/present/english/situation.ppt

MINAGRI, 2004. *Strategic plan for agricultural transformation in Rwanda.* Main document. Available online at URL: www.minagri.gov.rw/IMG/pdf/SPAT_ENGLISH_FINAL_DOCUMENT_1_.pdf

MINAGRI/CTB, 2005. *Etude d'identification. Programme d'appui au système national de vulgarisation agricole décentralisé.* Etude réalisée par Ted Schrader (KIT Amsterdam), Jean-Marie Byakweli (consultant Kigali) et Jean-Damascène Nyamwasa (PLANEEF Kigali). MINAGRI/CTB, Kigali, Rwanda.

MINALOC, 2005. *Territorial Administration. Administrative Units.* Available online at URL: www.minaloc.gov.rw/admin_territory/admin_entities.htm

OECD, 2006. *Rwanda.* African Economic Outlook. Available online at URL: www.oecd.org/dataoecd/33/55/36741760.pdf

Schrader, T., J-M. Byakweli, J-D. Nyamwasa, and G. Baltissen, 2006. *Action conjointe pour la prestation de services agricoles aux producteurs rwandais dans le contexte de la nouvelle décentralisation.* Working Document. KIT/Agriterra, Amsterdam/Arnhem, The Netherlands.

Notes

1. Note that a new bill on cooperatives is to be published soon. This bill will no longer recognize farmer associations outside of the cooperative legal framework.
2. INGABO also has another meaning in Kinyarwanda, namely shield (*bouclier* in French).
3. Since January 2006, these provinces have been reorganized into five regions.
4. *Gacaca* means a place to meet and discuss; it refers to the traditional place where dignitaries discussed village affairs and settled disputes.
5. Rwanda is divided into provinces *(Intara)*, districts *(Uturere)*, sectors *(Imirenge)* and cells *(Utugari)*. The district is the basic political-administrative unit of the country (MINALOC, 2005).
6. This equals about 1.83 USD.
7. *Gacaca* is also used by the Government of Rwanda for tradition-inspired participatory tribunals that were set up to deal with the 1994 genocide cases.
8. Data on surface areas of farm holdings have been collected by the authors. INGABO does not yet have a membership registration that allows for distinguishing member categories.

II MVIWATA's role in pro-poor service provision in Tanzania

Stephen Ruvuga, Richard Masandika and Willem Heemskerk

Introduction

Inclusive agricultural research and service provision

Agricultural research and extension systems in Tanzania continue to face the challenge to contribute to poverty alleviation through economic and agricultural growth. This growth in rural areas will not lead to development unless new and innovative ways of including everyone in this development process are introduced (URT, 2004a). The present Agricultural Sector Development Programme (ASDP), emphasizes demand-driven multi-service provider systems, which will include both public and private service providers, as well as civil society and farmers' organizations. In order to achieve this socially inclusive service provision, farmers' organizations need to be involved and lessons have to be learned from their experience in addressing the needs of the poorest (URT, 2004b).

Organization of farmers in Tanzania

The cooperative movement in Tanzania has always played a major role in marketing and price formation. The first Cooperative Ordinance was enacted in Tanganyika in 1931. Cooperative Unions and primary societies[1] were established as early as the 1930s, mainly for key cash crops, the most successful of which were the coffee and cotton cooperatives. Traditional leaders 'used' by the colonial government supported the formation of such primary societies. The unions became strong around independence time, but later came under political control before being fully disbanded and incorporated into the government structure (see Table 1).

The Villagization Act, which came into force in 1974, recognized a village as the co-operative unit in which each villager automatically became a member of the cooperative without paying entrance or membership fees. In effect there was no place

Table: 1: Overview of the development of farmers' organizations in Tanzania

1930s	Cooperative Ordinance Act, leading to the start of the cooperative movement
1960s	Political independence from the UK. Independent and strong unions, with assets such as ginning mills
1967	Arusha declaration, socialist economy, leading to control of the unions by the state
1974	Villagization Act. Villages become synonymous to cooperative units
1976	Cooperative Unions disbanded, assets confiscated, prices fixed by the government
1982	New Cooperative Act, leading to the re-establishment of the large state-controlled unions
1991	New Cooperative Act, allowing free organization of farmers
1993	Start of MVIWATA

Source: Modified after Chilongo, 2005.

for an autonomous cooperative society. In 1976 the Cooperative Unions and the primary societies were all disbanded by the government and reintroduced under the 1982 Cooperative Act. Large cotton and coffee unions, such as Shinyanga Regional Cooperative Union, Nyanza Cooperative Union, Kilimanjaro Native Cooperative Union, and Kagera Cooperative Union were re-established in 1984. The managers of these cooperatives were nominated by the state. As a result of these changes and the state control, the cooperative movement throughout the country lacked accountability, became dependent on state subsidies and hence economically unviable. Farmers lost confidence and trust in cooperatives, had no power and influence on the affairs and the running of the cooperatives. Unlike the past, when farmers were shareholders and managed their own affairs, these new cooperatives became only a network through which they could sell their crops. In response to this and the negative effects on economic development, the new Cooperative Act of 1991 was enacted to embrace the seven internationally accepted principles of cooperatives:
1. open and voluntary membership;
2. democratic member control;
3. member economic participation;
4. autonomy and independence;
5. education, training and information;
6. cooperation among cooperatives; and
7. concern for the community (ICC, 2006).

Further amendments resulted in new cooperatives to bring about the economic viability elements. The challenges brought about by the cooperative movement in the country led to the development of alternative farmers' organizations, which could be true representatives of farmers, and provide reliable advocacy. As a result, MVIWATA emerged in 1993 as a four-tiered network of farmer groups (Kaburire and Ruvuga, 2005). Its members include producer organizations such as dairy farmers and milk processors' associations, organizations in the commercial cash crop sector, but also the surviving viable primary cooperative societies or Cooperative Unions, Savings and Credit Associations and Savings and Credit Cooperative Societies.

Poverty and social diversity

Poverty is one of the key development challenges in Tanzania, which farmers are trying to address through collective action. Of the roughly 7 million households in Tanzania, almost 5 million (or 70%), live in rural areas. Of these rural households, 83% are headed by a person who works in farming or fishing, while others are involved in miscellaneous other economic activities. This leads to the conclusion that there are currently some 4 million farming households in Tanzania (URT, 2004a). Some 36% of Tanzanians fall below the basic needs poverty line and 19% below the food poverty line; in rural areas this is 39% of the population. Poverty remains overwhelmingly a rural problem, with 87% of the poor living in rural areas.

Three categories of smallholder farmers have been distinguished in Tanzania in the context of the Agricultural Sector Development Programme (URT, 2004a):
- *Most poor and least advantaged.* Households that fall below the poverty line, and earn, on average, an estimated USD 262 (Tshs 290,000) per household per year. This group includes the landless and casual labourers, households headed by females or orphans, and HIV/AIDS-affected small farm households. Such households are trapped in subsistence farming, a seasonal hunger gap, and inadequate cash income; typically cultivating less than 1.5 ha of land. Based on food poverty as an indicator it is estimated that 1.16 million households (29% of total farming households) belong to this category. The main intervention priorities are empowerment and emancipation through learning.
- *Poor.* Households below the basic-needs poverty line (USD 358 or Tshs 396,000 per year), have more diverse and less vulnerable livelihoods compared to the category of the 'most poor and least advantaged'. Their number is estimated at 1.21 million households, or 31% of the total number of farming households.
- *Better-off.* Households with access to small- or medium-sized areas (more than 3 hectares) of good land, which can play a positive role in extension, training and inclusion of their less well-off colleagues. Roughly 1.6 million rural households, or 40% of the total number of farming households, fall into this category.

This categorization illustrates the diversity of the farming population in Tanzania, although this is further compounded by geographic diversity. Local poverty profiles need to be used to monitor the proper representation of farmers in farmer groups and organizations. The ASDP foresees using local poverty profiles to monitor proper farmer representation in the District and Ward level Farmer Fora (FF). Overlying all the acknowledged factors of low productivity and profitability of small farms, is the excessive reliance on the labour of women, who are typically found to contribute more than 60% of their available work time (double that of men) to farming tasks, despite the burden of other family and household responsibilities. Poverty includes an important gender dimension. Female heads of households (25% of the total) earn less than 45% that of their male counterparts, while 69% of female heads of households live below the poverty line (URT, 2004a).

Presenting MVIWATA/MVIWAMO

MVIWATA is a national network of farmer groups in Tanzania. The organization was established in 1993 by small-scale farmers from the Morogoro, Iringa, Tanga, Mbeya and Dodoma regions (in the centre, southwest and northeast of Tanzania), who wanted to establish a farmer-to-farmer exchange forum. Sokoine University of Agriculture (through a Strengthening Communication Project at Morogoro) guided and facilitated its establishment, which finally led to formal registration of the organization with the government authorities in 1995. MVIWATA's mission is to link farmers' groups and local networks of such groups together into a sound and strong national farmer-based organization capable of ensuring representation and advocacy of their interests in decision-making processes at all levels, including through participatory training and communication strategies. MVIWATA's overall objective is to develop a strong and effective representation of farmers' interests by jointly confronting their needs and challenges, mainly concerning participatory communication, lobbying and advocacy, plus organizational strengthening to provide agronomic and marketing services. MVIWATA advocates strong organizations for smallholder farmers, establishing reliable markets for their farm produce, sustainable financial and technical advisory services, as well as empowered representation of farmers at all levels (Kaburire and Ruvuga, 2005).

MVIWATA was registered under the Society Ordinance Act in 1995 (No. 8612) and in 2000 as a Trust Fund. The Annual General Assembly (AGM), constituted by all members of MVIWATA, meets annually. This is the top decision-making level in the organization. Nomination of representatives takes place at local networks and the main criterion is that candidates must be individual members or cardholders of MVIWATA. The second decision-making body is the Steering Committee (SC), which consists of nine members. Members of the steering committee are elected every three years by the AGM. The Board of Trustees (seven board members), was established when MVIWATA was registered as a Trust Fund and is charged with providing the advisory function and custody of the organization's assets. The staff at national and intermediate levels implement the day-to-day activities of the organization. Intermediate-level networks are composed of farmers' networks at the district level, which comprise local networks at ward and village levels.

MWIWAMO is an intermediate-level network in Monduli District with four local networks. The organization was founded as a district-level intermediate network by 25 farmers from 12 farmer groups. The farmers were introduced to MVIWATA at a workshop organized in August 2001 and subsequently decided to start the process of forming a district network. MVIWAMO is a registered organization. It was legally registered as a district network of MVIWATA on March 17, 2004 (registration number SO 12374) under the Societies Ordinance. MVIWAMO became the first registered 'Intermediate Network' of MVIWATA. MVIWAMO aims to solve problems relating to:
1. market access, by means of cost-awareness training and market linkage strategies;
2. access to information, via community library and Internet facilities; and,
3. access to credits, through the formation and strengthening of 'Savings and Credit' groups.
4. It also addresses cross-cutting issues such as awareness on HIV/AIDS.

The organizational structure and decision-making procedures are laid down in the MVIWAMO's Constitution, supplemented by the Management and Financial Manuals. Elections for participation in Steering Committees are conducted by secret ballot during the AGM. Only members who have paid their annual fees and are fully-fledged members are entitled to vote or be elected to office. The procedures for elections are stipulated in the constitution as follows:
- The candidates lodge their nominations through their local networks but these networks do not screen the candidates.
- The AGM then elects seven members among the candidates, with at least one member from each local network.
- The elected seven members then compete for the posts of Chairperson, Deputy Chairperson and Treasurer.
- The only criterion for a candidate to be elected to one of these posts is that he/she receives a 'simple majority' of the votes.
- To achieve gender balance, at least one-third of the committee members should be female.

The SC stays in power for a term of two years. A member of the SC may be re-elected to the same post for two consecutive terms. An SC member may be elected for three consecutive terms if he/she moves to another post. Thereafter, the SC member may not be re-elected until at least one term has expired.

MVIWAMO mainly consists of small-scale farmers[2] but also accepts landless farmers as members, yet all must be members of a group first. The better-off farmers are encouraged to become non-voting associate members in order to assist and advise small-scale farmers. MVIWAMO is currently developing a new membership database, with members' identification and characteristics such as names, gender, sector, membership status, group, local network, skills possessed, and events in which a member participated (e.g., training, seminar, workshop, study tour, etc.). (Masandika and Mgangaluma, 2005; Masandika and Schouten, 2005.)

Membership of MVIWATA

MVIWATA covers the entire country, although membership and strength of groups and networks vary from one region to another. The constitution clearly states that the MVIWATA members should be small-scale farmers whose livelihoods mainly depend on agriculture. The reason is twofold: firstly, to avoid members who do not consider agriculture and associated activities as a priority for their livelihoods, and secondly, to avoid the chance that other professionals suppress the voice of farmers within their own organization.

The organization has members in all regions of Tanzania. Network size varies from 5-70 affiliated farmer groups, each with an average of 5-100 members. The targeted membership consists of 600,000 small-scale farmers, who can become members through their groups (presently 60,000) or as individual members; there are at least 10,000 cardholders (Kaburire and Ruvuga, 2005).

MVIWATA recognizes that social classes exist and that gender and age differences, when overlooked, may lead to exclusion from the development process. The principal groups affected are women, the young, and elderly people. The overall membership of MVIWATA has the following characteristics. The major source of members' income is from crop production (50%) or animal production (40%). Some 70% of the members cultivate less than three hectares. The main crop grown is maize (40%). Some 5% of the members are agricultural workers, although only 'farmers' officially qualify to become members. In 2001, 30% of the 3,000 individual members were female, while in 2003 this percentage had gone down to 18%, from a total of 5,200 members. The overall membership of women (combined group and individual membership) is estimated at between 33% (MVIWATA, 2004) and 45% (NAJK, 2005). Membership of MVIWATA is voluntary. Two categories of membership exist within the organization: ordinary members and associate members. Ordinary members include individual, groups and networks. Associate members are individuals, corporate bodies or organizations that may be solicited due to their outstanding contribution to MVIWATA. Individual members obtain their membership through their group; groups can register themselves as members and pay group fees. When a group becomes a member of MVIWATA, its members automatically also receive MVIWATA membership.

Networks at various levels (i.e., region, district, ward and village) may also become members. Basically, the networks comprise individual members of MVIWATA who reside in a particular location and converge together under the umbrella of MVIWATA. The 'individual members' have the constitutional right to vote and decide on matters related to the organization. They are responsible for paying membership fees and are the focal priority for events or activities organized by the organization. Previously, it was not necessary for all group members to join MVIWATA. There were various reasons why not all members of groups became members of MVIWATA, for example:

1. failure to be able to pay the necessary fees[3]. MVIWAMO has addressed this issue by allowing payments in instalments; and,
2. attitude issues: some prefer to wait until they see benefits being accrued by others, while others are not convinced of group action.

Members of MVIWAMO are individual cardholders of MVIWATA; the rate at which the number of these is growing depends on the mobilization, as well as the attraction of members to benefits and advantages they see from working in the network. MVIWAMO has 75 farmer groups with over 2,500 members of whom 450 (i.e., 18%) are also individual cardholders of MVIWATA. The group size varies according to the purpose of the group formation, with a minimum of five and maximum of 60 members (e.g., in the case of a women's group in a village who jointly owned a grain-milling machine). MVIWATA members are obliged to:

1. pay membership fees;
2. participate in meetings and activities;
3. manage group projects;
4. communicate and exchange knowledge and ideas between and amongst each other;
5. to vote, and to be elected into leadership;
6. prepare the groups' constitutions;
7. recruit new members; and
8. receive visitors.

Members enjoy certain benefits, which may include; training, exchange of ideas and knowledge, study tours, visitations, provision of capital, exposure and recognition. Expulsion can be caused by failure to pay annual subscription, going against the constitution or resignation. The process of expulsion starts with a warning, then suspension, followed by actual expulsion.

Gender dimensions

Profile

In the MVIWATA profile a number of indicators relating to gender mainstreaming were monitored in 2001 and 2003 (MVIWATA, 2004; Towo, 2004). See Table 2 for further details.

Table 2: Main MVIWATA gender mainstreaming indicators

Qualitative feature indicator	2001	2003
Expertise on gender issues	25%	48%
Gender as part of vision, mission and strategy	32%	60%
Formalization of gender policy in statutes, regulations and procedures	32%	60%
Gender as part of programmes and services	30%	60%
Role of women in policy- and decision-making	32%	52%
Possibility for women to gain full membership	88%	90%
Overall gender indicator	35%	54%

Although major progress has been made with the mainstreaming of gender issues, concerns remain about the role of women in leadership positions, as well as in relation to overall membership (fell from 30% to 18%, see above). However, the percentage of female board members has increased, from 30% in 2001 to 56% after the 2006 elections, due to specific attention given to this issue.

Equal opportunities

MVIWATA has installed mechanisms that ensure equal gender participation in the programme activities; this includes motivation, training, and provision of capital, initiation of study tours and other notable deliberate special considerations. These ensure equal benefits from the organization, which include; non-discrimination in constitutional provisions; equal opportunities in training and leadership; participatory planning in all stages of the decision-making process; and, specifically designed programmes that target women groups. Nevertheless some problems have been raised that relate to the lack of confidence among female members, aggravated by the lack of trust in women exhibited by some male members. This results in poor female participation in meetings and inadequate acquisition of skills through training and meetings. MVIWATA works to accommodate women, the young and the elderly in its activities – for example, women make up at least 40% of the farmers who attend training courses and farmer-to-farmer visits each year, and up to 65% of farmers benefiting from the credit fund are women and young farmers.

Decision-making

Constitutionally, one-third of the SC members must be female (the objective of this constitutional provision is to ensure that women participate in the decision-making body). MVIWATA's current national steering committee consists of five women and four men). The ratio in the appointed executive secretariat is however only 78:22. Although emphasis is given to equal opportunities, some special measures are still required to enhance female participation in the secretariat. The organization has further ensured women's participation through the following measures:
1. equal opportunities for all in elections; and,
2. special offers for women to attend meetings and workshops, including preferential leadership training.

A need was also identified for a review of the constitution to enhance greater women participation in the middle level management and leadership of the organization. Other strategies include; increasing the members' knowledge and awareness as well as encouraging participation. Encouraging women to form and/or join existing groups, as well as popularizing leadership training could further enhance improvement in this aspect. The female participation in the AGM has increased from 37% in 2003 to 43% in 2004, while the total attendance increased from 182 to 196. One-third of the 136 trained promoters in communication, entrepreneurship, advocacy and lobbying are women (Towo, 2004).

Advocacy

Women are heavily involved in agriculture, although in many cases mainly as labourers. However, things are now changing in many parts of the country as more and more women are beginning to take an active role in development activities. Be it politics, economy, health or social welfare, women are now taking centre stage in decision-making and thus assuming an important role in community development (Mfugale, 2005). Women are increasingly coming together to form all-female, common interest farmer groups, such as for vegetable growing and dairy goat enterprises, while their representation in mixed groups, including in office bearing oppositions, is significant and growing (URT, 2004b). Twelve years since MVIWATA started working with rural women members, the groups not only produce enough food for themselves but they have also turned to alternative sources of income. Female MVIWATA members have participated in the formulation of the national gender policy (Towo, 2004).

Building social capital

MVIWATA's main purpose is to develop strong and effective farmer representation in policymaking, service provision and markets, resulting in emphasis on organizational

> **Box 1: Examples of women's groups supported by MVIWATA**
>
> MVIWATA works with groups comprising both men and women, but the women are in the majority in the mixed groups. Women are chairpersons of the groups, others are secretaries and still others members of executive committees. Examples of women groups in Morogoro and Mvomero district of Morogoro region, which are supported by MVIWATA, are:
>
> *Sakina Abdallah* of Kinole village in Morogoro district heads a group that engages in drying fruits and vegetables (mangoes, pineapples, oranges, tangerines and breadfruit) by using solar energy. They undertake this activity after the usual daily chores. The group secures their market in hotels in Arusha and in Zanzibar, as there are no local markets. Despite the low production capacity, members of the group earn extra income that helps improve their quality of life.
>
> *Grace Mkwidu* is a leader of a group of women farmers in Mgeta, Mvomero district that propagates the use of organic pesticides, which are applied on vegetables, fruits and other crops. 'We encourage people to use the organic pesticides because they are not harmful to their health and not expensive'. A test was conducted to determine the effectiveness of both chemical and organic pesticides. The biggest advantage, according to the group leader, is that the vegetables or fruits can be consumed as soon as six hours after the application of the insecticide without affecting a person's health.
>
> *Francisca Lucas*, also from Mgeta, heads an all-women group that grows flowers to supplement the major farming activities in bananas, beans, cassava and vegetables. The flower business gave women an alternative to large-scale farming and generated substantial revenue. 'Sometimes it is not easy for women to haul sacks of beans or loads of cabbages, but it is easy to carry a bunch of flowers and the income is just as good,' MVIWATA's Shekilango clarified. The members of the group do not work full time on the flower farms, leaving them plenty of time to attend to the main farming activities.
>
> Source: Mfugale, 2005.

strengthening. MVIWATA therefore concentrates on the development of bonding social capital by the local and intermediate networks, thus also strengthening the bridging social capital, i.e., sensitizing group formation and networking. This provides an environment for efficient exchange of information among MVIWATA members, facilitating exchange of knowledge and experience within networks through meetings, workshops and joint projects. MVIWATA also aims to strengthen the linking social capital through local links with communities and other stakeholders from the public and the private sector. MVIWATA uses various approaches to build the capacities of its members. These include study tours, exchange visits and sensitization seminars. Short and long-term courses on land rights, trade and marketing, entrepreneurship, credit and savings, modern farming methods and cross-cutting issues, notably HIV/AIDS and gender. The need to invest in education as a way of addressing fundamental issues of productivity and efficiency in agriculture cannot be understated because the majority of MVIWATA-associated farmers are subsistence farmers with very low levels of education, which is an obstacle in the fight against poverty. Generally, MVIWAMO activities target the rural population at two levels:

1. at individual or household level through awareness raising and technical training, aiming at direct socioeconomic improvement in the households; and,
2. at organizational level by institutional strengthening of groups and local networks, leadership training, linking with other organizations, aiming at long-term empowerment of small scale farmers and pastoralists representation at all levels.

MVIWAMO aims to improve the socioeconomic situation of its members through:
1. institutional strengthening of groups and local networks;
2. facilitating the communication and exchange of knowledge and experiences between members, groups and networks among themselves, but also with other institutions; and,
3. as well as by empowering its members, also enabling them to choose and lead their own development.

Representativeness of MVIWATA

In many areas MVIWATA is strongly represented at the village level; small-scale farmers (irrespective of age, gender, farm type, religion, geographical origin or political opinions) can become members and hold responsible positions in the organization. MVIWATA has five organizational levels: individual members; local farmers groups; local networks; intermediate level networks; and the national level (with headquarters in Morogoro). MVIWATA operates under the motto 'Mtetezi wa Mkulima ni Mkulima Mwenyewe', which literally means 'defender of farmers' interests by farmers themselves'.

MVIWATA with its four-tier structure represents its members at all those levels, as well as through international networks.

The representation of farmers is strong at international and national levels (MVIWATA, 2004) and reasonably strong at the meso level (e.g., in the case of MVIWAMO), but the representation is considered weak at the district and local levels. The poorest farmers and women are seldom members of the organizations, as they are not able to invest time and resources in attending meetings. Their involvement, particularly in heterogeneous groups, is minimal, and their contribution to policy and lobbying activities is minor (Lema and Kapange, 2005,). In the ASDP programme Farmer Fora, will be organized at District and Ward level (DFFs and WFFs). Representatives with all kinds of social capital, such as farmer groups, cooperatives, women groups, seed-banks, etc. will get together in order to increase

Table 3: Level of representation of farmers by MVIWATA

Level	Farmers' organization	MVIWATA Links
International	IFAP, EAFF	Agriterra, IFAD
National	MVIWATA PELUM (regional network)	AMSDP, PADEP, INADES,
Zonal	e.g., MVIWAMO and other intermediate networks	EZCORE, Regional Consultative Council
District and Ward	Local networks, as well as district and ward Farmer Fora (FF)	Some projects, Monduli District MoU, District Network of NGOs and CBOs, Ward Development Councils
Community	Farmer Groups, Primary Societies, Associations	Groups can link up with other organizations

their control over the programme and eventually even receive funding for the agricultural advisory services provided (URT, 2004b). The WFF and DFF will have close links with Local Government Authorities. MVIWATA's grass-roots groups can, and intend to, become members of the aforementioned fora.

Role of farmers in MVIWATA

Farmers in MVIWATA are primarily members of the groups that are in turn members of local networks and middle level networks. Farmers can also become individual members, independently whether the group is a member. These members/cardholders can attend the AGM and elect leaders, approve annual reports and budgets, and strategic plans. Apart from this, members may have leadership functions at various levels. The members are also involved in a number of mainly voluntary services relating to both technical issues, as well as capacity development issues. The latter relate to both strengthening the existing groups and networks, as well as to establishing new groups and networks. The gender strategy foresees an equal participation of women in the AGM and all leadership and farmer promoter positions.

One major characteristic of the MVIWAMO leadership is that all members have equal chances of aspiring for a leadership position. In order to avoid particular members to dominate in leadership positions at various levels, members who hold leadership positions at lower levels (groups or local networks) are encouraged to allow other members a chance to aspire to positions in higher levels. Thus, being a leader at a lower level (e.g., a local group or local network) is not a prerequisite for being elected to the Steering Committee (SC) of MVIWAMO. MVIWAMO encourages other organizations to work with existing farmer groups to avoid groups to solely exist for the benefit of a particular project or NGO. MVIWAMO as a multicultural organization works with the traditional practitioners such as the LOIBON (Maasai traditional healers), LAIGWANI (Maasai traditional age leaders) as well as MABARAZA YA WAZEE (local fora for elders). MVIWAMO supports the ideas of organizations or 'movements', which fight against female genital mutilations, such as AFNET. MVIWAMO also has HIV/AIDS programmes, which aim to address traditional habits or values that have a role in spreading this disease.

SWOT analysis

MVIWATA's strengths, weakness, opportunities and threats were identified in exercises at different levels within the organization, but mainly through self-assessment.

The draft SWOT table was presented at a national workshop in January 2006, where it was updated and validated. The main observations will be further analyzed in paragraph 1.10.

Role of MVIWATA in access to services

The actual goal of MVIWATA is therefore to build a strong organization by mobilizing farmers to join groups in order to realize the objectives set by the farmers in initiating, implementing and monitoring their own economic and social plans. Farmers require appropriate services to support those plans and for that purpose farmers' organizations such as MVIWATA need to influence the type of services being provided, as well as to facilitate access to these services. These aspects get major attention in the new ASDP, through district and ward level FFs MVIWATA has yet to get fully involved in this evolving programme. However, MVIWATA has developed its own capacity in service provision. The organization, according to a cross-section of respondents, has successfully developed financial support for its members, and has organized training for farmers in accessing markets. Specific services of MVIWATA include:
1. initiating study tours and training for its members and leaders;
2. mobilizing farmers to join the organization at various levels and thereby increasing the number of groups and local networks;
3. training on opening and running rural banks (SACCOs) by and for members;
4. enhancing and improving the access to, and management of, rural markets;
5. publishing the 'Pambazuko' magazine on farmers' activities (8,000 copies distributed to members and partners); plus radio programmes.

Increasingly, MVIWATA and its intermediate-level networks are involved in production-chain-related activities and services, such as market and financial advisory services and projects.

Table 4: MVIWATA SWOT analysis for social inclusion and innovation

MVIWATA/ MVIWAMO	Strengths	Weaknesses	Opportunities	Threats
Constituency and membership	Inclusive member target group. Positive network feedback on poverty issues.	Limited coverage. Low networking capacity. Low social bonding capital strengthening for expansion.	Large potential constituency. Positive image farmer groups. Network contribution. Raised interest in farmer roles.	Many beneficiaries lack close contacts with the national level.
Legal status and legitimacy	MVIWATA is an autonomous, legally recognized organization.	Many Farmer Groups are not registered.		Agricultural policy with poor legal framework for role of farmers' organization.
Governance and democracy	Qualified board and management, regular meetings with members and mutual respect and trust.	Management committee is over-stretched. Poor gender strategy. Poor definition of leadership functions.	Building a strong grass-roots organization. Conducive policies for group and network formation.	Weak vertical social capital within farmers' organizations.
Purpose and objectives	Groups and networks with different poverty levels supported. Mobilization and formation of groups.	Inadequate M&E system. Support Staff overworked. Few linkages with partners at all levels.	Market policies increasingly pro-poor.	Poverty prevents full participation. Few farmers groups member. Donor dependence.
Institutional development	Strong links at national and international level.	Poorly defined interaction between the national and middle level.	Potential for linkages and networks at different levels.	Liberalization of financial markets.
Innovation development	National research and development links. Farmer Groups involved in new knowledge application.	No true partnerships at various levels. Poor links at the local level with research and advisory services providers.	The Farmer Fora (DFF, WFF) at various levels in ASDP.	Existing Farmer Groups (not leading to bridging social capital).

The assignment of policy and management responsibilities within the organization is such that, on average, there is no clear distinction between the secretariat and the leadership duties. This is a situation that needs attention in order to avoid duplication of efforts and interference. This relates to both the policy-making by the Board and the AGM and the management of service provision by the executive secretariat, as well as the influence on development policies by the Board and the facilitation of access to adequate services by the secretariat.

Concluding remarks

General

Research and advisory services have been addressing the needs of the smallholders, but the ASDP states that social inclusion needs to be improved and that institutional innovation will be required. Farmers' organizations, such as MVIWATA recognize that social classes exist and that gender and age differences, when overlooked, may lead to exclusion of certain groups in the development process. The principal groups affected apart from the poorest households are intra-household categories such as women, the young and the elderly. When implementing its activities MVIWATA attempts to accommodate such groups and make sure they are included. For example, women make up at least 40% of the farmers who attend MVIWATA-organized training courses and farmer-to-farmer visits each year. Up to 65% of the beneficiaries of a credit fund (for which MVIWATA facilitates access) are women and young farmers. In this regard, the aim is to enable such groups to develop confidence to manage their affairs, access common services including credit, and become active participants in decision-making and not simply remain passive members within MVIWATA. MVIWATA expects that this will indirectly reduce poverty and increase economic efficiency. It is also in MVIWATA's interests to work for all subsistence farmers, whether they are members of the organization or not. Demand by farmers to join the network is very high, which illustrates the perceived benefits, legitimacy and potential of the organization.

Inclusive farmers voice

MVIWATA focuses on a target group of 600,000 members, which is substantially less than the total number of smallholders in Tanzania (4 million households). The organization is of the view that, given its capacity, this is a realistic target group. The rural community in Tanzania is far from homogeneous; not all inhabitants are farmers; the number of agricultural workers in some regions is high; and, considerable differences are observed in poverty status between smallholders. MVIWATA considers the monitoring of the developing membership profile crucially important, in order to ensure adequate representation of the diversity of smallholders.

MVIWATA has a broad membership and officially draws its members from the farming population, but also has agricultural workers as members, although this is not a target group. However, the existing membership profile does not seem to take the observed diversity into account (MVIWATA, 2004). In addition, major gender differences exist, and many households are affected by HIV/AIDS. MVIWATA can have a significant number of female members, but the question remains as to whether these women adequately represent the estimated 25% female-headed households in rural Tanzania. The number of individual female members increased from 2001 to 2003, but total membership increased much faster. MVIWATA's fast growth is also due to the support for economic activities, such as production chain development, access to markets and financial services; it is possible that this attracts relatively more male than female members. Since MVIWATA has no detailed membership profile it is not known whether this is also happening with regard to the poorest farmers.

In any case, under the ASDP, FFs will be established at all levels, similar to the MVIWATA structure. This is both an opportunity for MVIWATA to set up these forums and influence decisions in agricultural development as well as a threat to MVIWATA, depending on the strength of the existing networks. Little mention is made in the ASDP programme of existing MVIWATA structures, which could lead to the conclusion that a view exists that MVIWATA is not adequately representing all farmers in some areas (URT, 2004b). ASDP has started to make district- and ward-level inventories of all existing farmer groups and these will be invited to have representatives in the FFs. The FFs will have important roles in setting the research and advisory services agenda, facilitating access to services and, eventually, the procurement of these services.

MVIWATA's agenda is increasingly dominated by themes such as access to market and credit services, which are important but not automatically pro-poor. How can MVIWATA members address the demand for services and make the access to these more inclusive? In other words, if the service provision is not inclusive then demand articulation and farmer empowerment will require improvement. MVIWATA's links with services at the local level, to the extent that farmer groups and networks have a real influence, are currently limited. MVIWATA has a stronger lobbying and policy influencing voice at the national and international level. Although MVIWATA has issues such as food security and HIV/AIDS on its agenda, does the lobbying for attention and support for these themes receive sufficient attention if many of their members also have other priorities?

Inclusive service provision

MVIWATA, and the networks at intermediate and local level, are increasingly involved in advisory services for their members as well as for others through contracts with the public and private sector. This development will be supported under ASDP, since farmers' organizations can also be contracted using public funds. MVIWATA needs to develop a policy on how to work with the District and Ward FFs. DFFs and WFFs can provide an opportunity for MVIWATA to strengthen its local and intermediate networks, while the DFFs and WFFs will also become involved in contracting MVIWATA for service provision. The present relationship between WFFs and existing farmers' organizations (such as MVIWATA) is not clear, also since a parallel structure appears to be forming with WFFs, DFFs and National FFs. MVIWATA also needs to develop a policy in terms of pro-poor research and advisory service provision for innovation in general. For example, how will special attention be given to represent and increase membership by the poor, such as female-headed households, HIV/AIDS-affected families and other disadvantaged groups?

Future developments

Possibly because ASDP has insufficient confidence in the inclusiveness of MVIWATA, the programme has developed a number of eligibility criteria for inclusive participation in the agricultural sector programme through the DFFs and WFFs (URT, 2004a). These are:
1. membership of each farmer group or forum should include at least 70% smallholders[4];
2. priority will be given to districts and wards with a high incidence of poverty[5];
3. targeting by prescribing a minimum of 40% women in all groups and programmes with disadvantaged (landless, HIV/AIDS affected) at least at 10%, an equitable balance in farmer resource persons, and attention to specific priorities for the most disadvantaged; and,
4. mechanisms to guarantee access to goods and services by the rural poor.

These targets for ASDP are considered to be a good basis for MVIWATA's pro-poor innovation development strategy.

References

Chilongo, Thabbie, 2005. *Tanzanian Agricultural Co-Operatives: An Overview.* A Draft Report. Moshi University College of Co-operative and Business Studies, Moshi, Tanzania, September 2005

Chilongo, Thabbie, 2005. *Tanzanian Agricultural Co-Operatives: An Overview.* A Draft Report. Moshi University College of Co-operative and Business Studies, Moshi, Tanzania, September 2005.

ICC, 2006. *Cooperative principles of the International Cooperative Alliance.* Inter-Cooperative Council http://www.icc.coop/learn/about/documents/COOPERATIVEPRINCIPLES.pdf

Kaburire, L., and S. Ruvuga, 2005. Networking for agricultural innovation. The MVIWATA national network of farmer groups in Tanzania. In: Wennink and Heemskerk, 2005. *Farmers' organizations and agricultural innovation. Case studies from Sub-Saharan Africa.* Bulletin 374. Royal Tropical Institute, the Netherlands.

Lema, N.M., and B.W. Kapange, 2005. Farmers' organizations and agricultural innovation in Tanzania. The sector policy for real farmer empowerment. In: Wennink and Heemskerk, 2005. *Farmers' organizations and agricultural innovation. Case studies from Sub-Saharan Africa.* Bulletin 374. Royal Tropical Institute, The Netherlands

Masandika, R., and A. Mgangaluma, 2005. Linking farmer groups with various agricultural service providers. The MVIWAMO District network of farmer groups in Tanzania. In: Wennink and Heemskerk, 2005. *Farmers' organizations and agricultural innovation. Case studies from Sub-Saharan Africa.* Bulletin 374. Royal Tropical Institute, The Netherlands

Masandika, R., and C. Schouten, 2005. Farmer inclusion and agricultural innovation. Case of MVIWATA Monduli, MVIWAMO. Tanzania.

Mfugale, Deodatus, 2005. Farmers groups benefit women. *The Guardian* 2005-08-13, Tanzania. Available online at URL: www.ippmedia.com/ipp/guardian/2005/08/13/46919.html

MVIWATA, 2004. *Organizational profile*, April 2004, Tanzania.

NAJK, 2005. Available online at URL: www.agro-info.nl/scripts/org_info.asp?lang=nl&sender=m&org=mv&tiep=1

Towo, Esther, 2004. *The Gender Dimension of Rural Producer Organizations in Tanzania.* Working Paper 2004:131. Norwegian Institute for Urban and Regional Research. Available online at URL: www.nibr.no

URT, 2004a. *Agricultural Sector Development Programme. Agricultural Services Support Programme.* Programme Document and IFAD Appraisal report. Volume I: Main report, Mainland; Volume II: Zanzibar Appraisal Report; Volume III: Core Programme Document, Final Draft.

URT, 2004b. *Farmer Empowerment Programme Component.* ASDP/ASSP Working Paper. Revised version 12, March 2004.

Notes

1. Village-level cooperatives, the building blocks of the Cooperative Unions.
2. Farmers are people engaged in agricultural production by tilling land, keeping livestock, bee-keeping or fishing.
3. Individual entry fee is Tshs 1,000 and the annual subscription = Tshs 2,000, while this is Tshs 2,000 and Tshs 10,000 for farmer groups and Tshs 20,000 annual fee for local networks.
4. Smallholder farmers in Tanzania cultivate less than 5 ha rain-fed or less than 1 ha irrigated land, no limits to livestock husbandry in order not to exclude pastoralists.
5. The formula used is based on food poverty head count weights with a factor of 0.2.

III UCPC's role in pro-poor service provision in Benin

Clarisse Tama-Imorou, Bertus Wennink, and E. Suzanne Nederlof

Introduction

Cotton is the most important cash crop in Benin: it contributes about 10% to the GDP of the country through an estimated 90% of the value of agricultural exports, and it accounts for an important share in farmer household incomes (Minot and Daniels, 2002: pp. 9 and 18); poverty is therefore lower among cotton growing households than among the rural population at large (Ibid: p. 19). Cotton is being grown by all categories of farm households although cotton growers tend to have larger farms than non-cotton growers (Ibid: p. 18). Since the 1980s cotton production has rapidly increased, mainly through area expansion. One success factor is the creation, with support from state services, of producer groups that are responsible for handling input supply and marketing of cotton at the village level.

Until recently the cotton sector was entirely controlled by public sector stakeholders; the main players were: a parastatal in charge of input distribution and marketing (SONAPRA), the national cotton research centre CRA-CF and the agricultural extension service CARDER (Sinzogan *et al.*, 2006). Problems of inefficiency and mismanagement forced the Government of Benin to reform the sector in order to maintain and improve the competitive position of Benin cotton. However, the integrated production chain approach *(filière intégrée)* was maintained, whereby chain operations and support activities are linked and coordinated, as well as a single 'floor/basic price'[1] maintained across the entire country.

Since the beginning of the 1990s the cotton sector has been involved in a major liberalization and privatization exercise: gradual withdrawal of the state from input supply and marketing of cotton, the responsibility for which is being transferred to the private sector; reorientation of state services on regulatory responsibilities and so-called 'critical' functions (e.g., research and extension); and, an increasing role for producer organizations in governing the sector including for price setting. This is all expected to lead to enhanced transparency and efficiency, as well as a greater share of producers in the prices paid for cotton. However, the combination of declining prices on the world market and problems while implementing the reforms have led to a decrease in cotton production over the last few years (World Bank, 2005: pp. 86-93).

As a result of the reforms the cotton sector stakeholders are no longer simply linked through a chain approach but rather through a network approach; important in this transformation has been the creation in 1999 of the inter-professional association (AIC) as a private sector platform for consultation, and representation of cotton sector stakeholders. Already in 1994 the producers initiated their own network of organizations: GVs at village level, UCPs at district level, UDPs at departmental level, and FUPRO at the national level (Kouton et al., 2006). This gave producers a strong position vis-à-vis other stakeholders. However, management problems within the sector made some producer groups create their own apex organizations in so-called 'break-away' networks (Sinzogan et al., 2006: p. 48; see Table 1).

Table 1: Actors in the formal and 'break-away' cotton networks in Benin (by stakeholder category)

Category of stakeholders	Definition	Networks	
		Formal	Break-away
Primary stakeholders	Those who are directly affected.	Producers (GVs and UCPs).	Producers (GVs and UCPs).
Intermediate stakeholders	Intermediaries in delivery or execution of research, resource flows and activities.	AIC, ADIAB, CSPR, CAGIA, UDPs and FUPRO, banks, CRA-CF, input suppliers and ginners.	AGROP, FENAPRA, input suppliers and ginners.
Key stakeholders	Those with the power to influence or 'kill' an activity.	Government, AIC, CSPR, CAGIA, FUPRO, ginners, bank.	Government, AGROP, FENAPRA, input suppliers and ginners.

Adapted from Sinzogan et al., 2006.

The break-away organizations are especially present in the most important cotton production areas in the north and centre of Benin. FUPRO still remains the main producers' network and is generally considered to be trustworthy. In 2005 FUPRO's member organizations implemented the latest reform: they sub-divided according to cotton and other cash crops such as rice. Although producer organizations already largely depended on cotton for their functioning, until 2005 they still formally also handled inputs and marketed products for all other cash crops. At present, the UDPC and UCPC (note the added 'C' for cotton) therefore only comprise cotton producers, and the GVPC replaced the GV. Their new, national apex organization (ANPC) is still a member of FUPRO. The focus on cotton was also motivated by resolving a 'free rider' problem, which involved farmers who didn't actually grow cotton but obtained inputs by promising to grow cotton and to have the inputs deducted when marketing cotton through their groups. However, many farmers used these inputs for crops other than cotton, but their debts nevertheless had to be reimbursed by the group due to the collective repayment obligation *(caution solidaire)* (Sinzogan *et al.*, 2006).

The history of the village producer groups is closely linked to that of the agricultural extension service (formerly known as CARDER, now called CeRPA), which established the groups and, until the creation of unions, also supervised them. CeRPA is a deconcentrated public service that is present in all districts of Benin. Other agricultural service providers are Non-Governmental Organizations (NGOs), which are often involved in the implementation of donor-funded rural development projects (see for example the case study on ACooBéPA elsewhere in this publication) and private sector enterprises (e.g., input suppliers). Still the agricultural extension service remains by far the most important service provider for cotton producers. Cotton research is managed by the cotton research centre, which has two branch offices located in northern and southern Benin. Cotton research is one of the specialized programmes of the national institute for agricultural research (INRAB), which also runs three eco-regional programmes, in southern, central and northern Benin; the latter are managed by regional research centres (Kouton *et al.*, 2006). Finally, the network of rural banks (CLCAM) is the main credit facility for cotton producers; this is made possible because floor/basic prices for cotton are guaranteed.

This case study presents the *Unions Communales des Producteurs de Coton* (UCPCs) which are member organizations of FUPRO at the district level, in three different districts: Kandi (in the northern Alibori region of Benin), Boukoumbé (in the northeastern Atacora region) and Djida (in the central Zou region). In both Kandi and Boukoumbé, agriculture is the main activity and cotton is the core cash crop, despite the crisis the cotton sector is currently suffering. In Djidja, agriculture is also the main activity but cotton production is a rather recent phenomenon. Data and information presented in this case study were gathered through semi-structured interviews with farmers, collective interviews with groups, focus group discussions, observations, informal chats and through a literature review.

Presenting the UCPC

The UCPCs (before 2003 they were known as USPPs[2]) were created as apex organizations of village producer groups (GVPC). The main functions of village producer groups are economical as well as social:
1. to handle input supply and marketing of cotton; and,

2. to contribute to community development (see Box 1).

In return, unions support the functioning of member groups (e.g., coordination of input supply and marketing of cotton, technical assistance and financial control), participate in local development and represent members at other levels.

> **Box 1: Functions of the village producer groups**
>
> - handle input supply and marketing of cotton on behalf of their members;
> - provide a solidarity intention to the individual members to receive credit;
> - organize education and training of members for cooperative management;
> - contribute to the building of community infrastructure;
> - promote village community development;
> - facilitate mutual and solidarity practices among members.

The reasons behind producers creating their own unions included the eagerness to manage their own activities, the state's withdrawal from support activities, and the lack of trust farmers had in agricultural extension agents when supervising producer groups. The common factor linking the producers is their willingness to exchange ideas and find solutions to the problems they face. The creation of district unions was also part of the agricultural reforms introduced by the government at the beginning of the 1990s (see for example the typical statutes and by-laws that have been elaborated and disseminated; Anon., 1995). Both the village producers' groups and the UCPC are cooperative organizations as defined by the cooperative legislation of Benin. A number of UCPCs are registered by the agricultural extension service but do not have yet official recognition. Kandi UCPC, for example, has been registered at the CeRPA but does not yet have an official status. This is due to the fact that the introduction of an exclusive focus on cotton in 2005 required an adaptation of statutes and by-laws that has not yet been completed. Nevertheless, the UCPC continues to function as a cooperative union and represents its member organizations at apex entities such as FUPRO.

According to the people interviewed, Kandi UCPC was created in 1985 (as a USPP). The union is concerned with one of the most important cotton production areas with an annual production that fluctuates between 35,000 and 45,000 tonnes (data for 2001-2005; SNV Conseils, 2005). However, it should be emphasized that, since 2003, about a quarter of the district's cotton production is marketed by break-away networks (Ibid).

Boukoumbé UCPC originated (also as a USPP) at the end of the 1990s during the process of organizing village cotton producer groups at the district level. Up to the end of the 1980s these village groups had been organized around groundnut production and benefited from state subsidies for input supply and marketing. When such subsidies were terminated, farmers turned to growing cotton, despite the area's relatively unfavourable rainfall and soil conditions, which also explains the relative low volume of cotton produced in this district, not exceeding 600 tonnes per year (Wennink and Dotia, 2004: p. 4).

Although the Djidja UCPC has existed since 1985 (as a USPP), cotton production is recent in Djidja district and never reached more than 14,000 tonnes per year; this has stagnated at around 8,000 tonnes per year (data for 2001-2005; SNV Conseils, 2006). Two main reasons are forwarded by Djidja union members for this decline:
1. the lack of rainfall; and,
2. the resignation of group member producers who no longer want to be obliged to pay for debts contracted by other cotton producers (Ibid).

The amount of cotton produced is important when considering the financial resources of an UCPC. Cotton levies are by far the most crucial income source and are used to reward the village groups and the district unions for handling operations in the cotton chain.[3] Furthermore, village groups receive rebates *(ristournes)*[4] when final prices of cotton that has been sold turn out to be higher than what was expected and agreed at the start of the planting season. Other financial resources for a UCPC include:
1. 'social shares'[5] *(part social)* paid by member groups;
2. membership fees; and,
3. income-generating activities such as renting trucks to transport cotton.

The level of both the social share and membership fees is decided by the UCPC's general assembly.

UCPCs are governed by 'Boards' *(conseil d'administration)* whose members (up to 20) are elected by the general assembly *(assemblée générale)* of representatives from the union's member organizations. The Board comprises a core executive board *(bureau exécutif)*, which in turn is monitored by an auditing committee *(comité de contrôle)*. According to the by-laws each member GVPC of a union appoints three representatives to attend the annual general assembly with a mandate to elect board members. All UCPCs have a salaried technical staff consisting of an executive director, an accountant and a secretary, and some support staff (e.g., drivers), though this varies from one union to another.

Membership of the UCPC

Village cotton producer groups (not individual farmers) are the UCPC members. Before splitting up village producer groups (GV) into cotton producer groups (GVPC) and other groups, every producer, whether man or woman, could become a group member as long as he/she operated within the area of activities as defined in the group's statute, and paid the requested fees. Therefore, at that time, members of a village producer group included livestock owners, fishermen, producers of different crops such as cashew growers, and even women involved in processing activities. Some of these groups, including women's groups, created their own village organizations and registered as such with the agricultural extension service. They could also become a member organization of the district producers union UCP (without the C).

Today, in order to become an individual member of a village cotton producer group (GVPC), a farmer needs to grow cotton, and membership of the UCPC is exclusively reserved for these cotton producer groups. This has an impact on the membership base, particularly in those regions that are less appropriate for growing cotton, such as the Boukoumbé and Djidja districts (see Table 2). In these districts, the membership base of the UCPCs has significantly narrowed and, more importantly, decision-making concerning the utilization of funds generated through cotton production is now formally being made exclusively by the cotton producers. In Kandi district a very large majority of smallholders grows cotton, in combination with other crops and activities. As a result of these developments, in all districts, membership of women's groups has probably diminished, since men traditionally dominate cotton production, and women's processing groups can no longer belong to the UCPC.

Table 2: Overview of the membership base of the UCPCs

	UCPC Kandi	UCPC Boukoumbé	UCPC Djidja
Inhabitants	n.a.	n.a.	n.a.
Villages	42	n.a.	n.a.
GVPCs	91-97	52	84
Members	13,400	3,700	n.a.
Members – men	10,390	n.a.	n.a.
Members – women	2,960	n.a.	n.a.

n.a.: not available.

Formal criteria for a village cotton producers' group to become a member of a UCPC include:
1. paying a social share and membership fees;
2. providing a written statement to agree to the UCPC rules and regulations; and,
3. being registered with the extension service.

However, different players negotiate informal rules concerning, for example, procedures to be adhered to and the fees to be paid. A village cotton producers group submits a written request to join the union's general assembly; in order to be admitted, the group has to be judged 'functional and viable' (e.g., organize regular meetings, keep proper accounts, manage in a transparent way, and conduct profitable activities).

Although UCPCs do not have records on individual members, interviewees mention the volume of cotton produced and ethnic groups as criteria used for distinguishing between different village groups and members with regard to qualifications for leadership. Table 3 presents an overview of the main ethnic groups in the three districts. As one interviewee stated: 'During board meetings, one can easily identify which ethnic group members come from, if through nothing else than the way they talk or behave'.

Table 3: Representation of ethnicity in the UCPCs

UCPC Kandi		UCPC Boukoumbé		UCPC Djida	
Ethnicity in villages	Ethnicity in UCPC	Ethnicity in villages	Ethnicity in UCPC	Ethnicity in villages	Ethnicity in UCPC
Bariba*	Bariba*	Ditamari	Ditamari*	Fon	Fon*
Mokolé*	Mokolé*	Yandé	Yandé*	Nagot	
Dendi*	Dendi*	Lamba	Lamba	Adja	
Boo	Boo	Lokpa		Yoruba	
Fulani	Fulani/Gando	Fon			
Migrants (Atacora)		Yoruba			
Fons		Djerma			
Yoruba		Fulani			
Djerma					

NB: The ethnic groups that are represented on the UCPC Boards are indicated with a star.

In Kandi district, three ethnic groups dominate the membership, with one (the *Bariba*) being the majority group. Members are not only distinguished according to ethnicity but also according to their spatial position within the district: those living in the centre (Kandi town) or at the periphery (villages). According to the interviewees this last criterion has its origin in the struggle for political power in the district.

Several ethnic groups inhabit the villages in the Boukoumbé district. However, they are not all represented by the UCPC for the simple reason that only a few ethnic groups (the *Ditamari, Yandé* and *Lamba*) are involved in growing cotton (mostly because only specific parts of the district are appropriate for cotton cultivation). In Boukoumbé the UCPC has an explicit policy of combating financial mismanagement: UCPC supports GVPCs to identify 'bad payers' (producers who do not reimburse their credit) by sending an inquiry team that 'forces' such farmers to pay.

Only three ethnic groups are present in the Djidja district UCPC, with one being the majority group (the *Fon*). In this district, there also seems to be an antagonism between the centre and the periphery. Until recently, cotton producers residing in Djidja town held all posts on the Board. The other members contested this and, during an important meeting, it was decided that the Djidja district would be divided in three sections, which would be individually represented during the elections of board members.

Gender dimensions

Although cotton production is seen by many as a 'man's business', women are also participating. Women heads of farmer households produce cotton on land that they bought or inherited from their husbands. They are registered as such under their own name with the village groups (GVPCs). This group of female-headed households is rather small since inheritance of land by women is rare, and 'free' (not married) women are culturally not well-accepted. However, most women in the districts concerned are workers in cotton production who operate under the authority of their husbands and they are not individually registered with the village groups (Kamminga, 2005).

The national federation of cotton producers' unions (FUPRO) has an informal policy to allot 25% of the elected representative positions to women. FUPRO's general assembly consists of 48 members (eight from each departmental union) including 13 women; however, the FUPRO Board of Directors includes 12 men and only one woman (2005). The quota system has neither been formalized nor been accompanied by adequate capacity building programmes for women members to enhance skills such as leadership or managing group dynamics. Women's groups that undertake economic activities collectively (other than in cotton production and marketing) are considered by FUPRO staff to provide an excellent learning opportunity for voicing women's needs and defending their interests (Ibid). However, such opportunities diminished considerably after the decision to focus membership criteria of unions on cotton production. Lack of women candidates and opposition by men are forwarded as the main reasons for the poor representation of women on Boards and this is confirmed by the situation in the three UCPCs studied.

Women in Kandi and Djidja are not represented on the UCPC's Board of Directors. The Kandi UCPC has never had a woman board member, despite the fact that the FUPRO quota system was recommended to voters during elections. Reasons brought forward include the fact that husbands do not allow their wives to join and the perception that women do not have the required knowledge and skills for this type of work. In addition, an informal rule excluding women from being on the Board is that to become a board member you have to be a large producer of cotton (several tons); women usually grow only small areas of cotton.

Only Boukoumbé has women members of the UCPC Board. In this district, board members other than those on the executive board have a specific task within the Board (see above), such as information and communications manager, input manager, marketing manager, training manager, advisor, etc. The two women board members are the deputy treasurer and the social affairs manager. One interviewee explains that 'the deputy treasurer can be counted amongst the men: she is a literate person, a large cotton producer, and appreciated being made responsible for her GVPC'. Also, she operates in a (donor-supported) group that strives to enhance women's voices in governance institutions in rural areas. Another interviewee explains that: 'women are board members because development partners have advised them to do so and we want to keep them (the donors) happy'.

Building social capital

Since the creation of village producers groups (GVs) was a national policy, these groups are found all over Benin in almost all villages. They were created independently from cotton production but this cash crop turned out to be the 'lifeline' for most GVs. Membership criteria of the former village groups were rather flexible since one needed to just be a 'producer', and the levels of fees were defined by the group's general assembly. With the transformation from GVs to village cotton producers groups (GVPCs), 'producing cotton' became a more stringent membership criterion and encouraged other (cash crop) producers to form their own groups. However, particularly in those areas that are suited to cotton production, the majority of farmers grow cotton in addition to other crops.

Farmers are keen to produce cotton because it is a cash crop providing them with liquidity and because it facilitates access to inputs and credit that are also used for crops other than cotton. Whatever the size of the cotton plantation, as long as a person pays the fees and subscribes to the collective repayment obligation, then they can become a member. Nevertheless, social cohesion of village cotton producers groups is seriously threatened by the growing debts, because there are members who sell inputs or use them on other crops and therefore are not able to repay their debts. This phenomenon is expanding because of the decreasing prices of cotton. Conflicts within member village groups are therefore by far the most recurrent issue during UCPC Board meetings.

All cotton producer groups are community-based with ordinary members, as well as board members, coming from the same village. Until now, these village groups (GVPCs) were still the privileged farmers representatives for the agricultural extension service, since the groups were mostly formed at the initiative of (and with support from) the extension service. The splitting up of village groups, as well as the growing support for agricultural diversification by donor-funded projects (e.g., concerning rice, cassava, yam, and Irish potatoes), has resulted in the emergence of other village producers groups who in turn are the privileged farmer representatives for these new projects (Kouton *et al.*, 2006). However, village cotton producers groups continue to play a central role in the life of their villages. A guaranteed price for cotton and rebates (related to favourable world market prices and high quality cotton) provide them with important financial resources that are being used for developing community infrastructure, such as schools and health centres. Furthermore, village producers groups (mainly involved with cotton) have benefited for years from literacy training and management support, which means that former board members of these groups are often solicited to lead other community-based groups.

Under the former state governance of the cotton sector, relationships between cotton producers' organizations and other stakeholders in the sector were rather limited, since all cotton production and processing-related activities were managed and controlled by the parastatals. Since the liberalization and privatization of the cotton sector, cotton producer organizations (and particularly the UCPCs), are increasingly responsible for managing relations with:
- Private input supply firms, cotton transporters and ginners.
- The national agency (CSPR) that monitors cash flows for input procurement, payment of cotton and reimbursements.
- Agricultural services such as the district extension services (CeCPA), local credit providers (CLCAM), and local development projects.

UCPCs also represent cotton producers:
- Vis-à-vis district authorities and services.
- In platforms at the sub-national level through UDPCs (e.g., in priority setting for agricultural research; see Gotoechan-Hodonou *et al.*, 2005; Kouton *et al.*, 2006).
- In the national cooperative (CAGIA), through FUPRO, for input procurement.
- At the AIC level, through FUPRO, for negotiating cotton prices, funding of cotton research and agricultural extension, and planning of production and marketing.

Input providers are currently actively courting unions to gain a share of the input market because procurement of fertilizers and pesticides is handled by a cotton producer-led cooperative (CAGIA); see Table 1. Input providers also promote their products through their commercial representatives, who advise farmers and organize training sessions for cotton producers. Some UCPCs are also being approached by ginners to sell their cotton to them directly, despite AIC agreements about allocation of the planned production to private ginners. This is due to the fact that overall ginning capacity is greater than the production capacity, which leads to under-exploitation of ginning mills (Sogbohossou *et al.*, 2005: p. 92). As a result of these recent developments, cotton producers unions face many other problems beyond their own organizational capacities, such as: producers receiving their inputs from one network and selling their cotton to another; producers selling the inputs they received at reduced prices for short-term financial gain; provision of bad quality inputs; delay in both input provision and payment of cotton provided; transport irregularities, etc. This puts an enormous strain on the relationships between the different cotton producers' organizations, as demonstrated by the emergence of the 'break-away' networks.

In 2003 the Benin decentralization policy became effective, by devolving decision-making on local development to elected district authorities. However, the resources needed for the growing number of tasks did not keep pace with the increasing assignment of decision-making powers. This strengthened the already crucial role of UCPCs in district development. On the one hand the unions further developed their relationships with the district agricultural extension services, which now had the formal mission to contribute to local, economic development. Client-provider relations are also enhanced through (partial) funding of extension services by AIC through cotton levies (Sogbohossou *et al.*, 2006). On the other hand, UCPCs are submitted to growing political pressure by local authorities to contribute to local development that goes beyond the usual funding of community infrastructure. This is particularly the case in the heart of the cotton producing areas of Benin, such as the Kandi district.

Unions also look beyond the boundaries of the district, as in Kandi. The Kandi UCPC provides scholarships to members' children and contributes to building residences for students in the university campuses in Parakou and Cotonou. It considers this to be an important investment in the capacity of future staff of public and private sector organizations on which they will someday rely. Similar to the development at the village level (GVPC), UCPC board members are also elected in district community organizations because of their skills and network contacts. Several UCPC board members in Boukoumbé are also heads of sub-districts *(arrondissements)*.

Representativeness of the UCPC

Each GVPC member organization has three representatives in the general assembly that elects UCPC board members. Candidates must comply with criteria that relate to issues such as being a cotton producer and being of good 'moral' conduct (see Box 2). In most cases the Board has up to 20 members with a core Executive Board of five members: the president, the vice president, the secretary, the deputy secretary and the treasurer. The first criterion to become an executive board member seems to relate to the level of production, the most important cotton producer usually becomes the president: 'To become president, you have be an important producer, who knows what it is to grow cotton and the problems that are encountered. It's a question of practice and not a diploma'. A reason behind the major cotton producer becoming the president is also that members seem to hope that this person will then be able to pre-finance some of the UCPC's activities: 'Our president pre-funds some of our trainings. He is rich. How could we do this otherwise without such a possibility?'. Therefore a second, but linked, criterion is related to the producer's financial assets.

Box 2:   Official criteria for becoming a UCPC board member

- Beninese nationality
- not being convicted of any crime
- not participating in any activity competing with UCPC objectives
- being recognized as a settled cotton producer at the grass-roots level
- being engaged in cotton production for at least two years and staying in cotton production during mandate
- being of good morality and being available
- not participated in theft, embezzlement, etc.

In general, reading and writing in French are other important criteria to becoming an executive board member, not only for the secretary but also for the president. 'Political capital' also seems to play a role. In addition, proof of honesty and trustworthiness during past assignments (as sub-district chief or on a GVPC Board) is important. Social relations and control seem to play a major role in the elections: 'We know each other and each others' relatives and history, so we also know who can do the job'.

In Kandi the five posts on the UCPC Executive Board are divided amongst three dominant ethnic groups (see Table 2). Reasons for their dominance are, according to the interviewees, the fact that the others are not traditional residents and are considered 'strangers', are a minority, spatially dispersed, and 'barely educated'. Another type of differentiation is amongst 'those from the centre' and those from the 'periphery'. The Fulani, semi-nomad livestock holders, who rarely received any formal training, revolted against this type of division, but were told that they could identify themselves with some of the other ethnic groups. Furthermore, the three dominant groups decided upon a rotational leadership to be followed: two groups alternate as president (a *Bariba* or *Mokolê*) and the vice presidency is always from the third group (a Dendi). The other three posts are occupied by representatives from the other two groups (either a *Dendi* or a *Bariba*).

In Boukoumbé the two major ethnical groups (the *Ditamari* and *Yandê*) divide the Board positions between themselves because they are greater in number and originate from the Boukoumbé district (see Table 2). These two majority groups also understand the same language *(Ditamari)*. The Boukoumbé Board is a mixture of illiterate and literate members as well as persons with other 'higher' positions, such as sub-district chiefs and retired extension agents. Furthermore, the UCPC does not have its own office building and holds its meetings in the offices of the district agricultural extension service.

In Djidja a sole ethnic group (the *Fon*) occupies all posts on the UCPC Board (see Table 2). Members consider the president to be a rich man: he is an 'agricultural entrepreneur who owns a large farm and employs salaried workers'. The history of the Djdidja UCPC Board is characterized by cases of financial mismanagement and embezzlement of funds *('malversations')* by both elected officials and staff members, and as a result, several board members have been forced to resign and the others now combine different posts.

Although election procedures and eligibility criteria are formalized in statutes and by-laws, other mechanisms have gradually been institutionalized in all three cases. First of all, the number of representatives from GVPC member organizations is being rectified on the basis of the volume of cotton produced by a village group. This criterion is also considered when electing individual executive board members. Secondly, the three GVPC representatives are not always simply elected; they are often restricted to three self-designated representatives: the president, secretary and treasurer of the GVPC. Finally, added considerations, such as ethnicity and residency mean that, strictly speaking, free democratic elections are not being held. Ethnicity in itself is not the issue, but minority groups are often less well-trained (e.g., semi-nomadic livestock holders) or don't have secure land rights (e.g., migrants from other areas). All UCPCs have introduced rotational board memberships in order to strike a balance between different areas and groups within the district.

Role of farmers in the UCPC

UCPCs have a key position in the management design of the cotton sector in Benin: they are considered the 'hinge' between the national level (where decisions are taken on input supply, cotton prices and marketing schemes), and the local level (where logistics are organized). Therefore, the UCPC remains the main platform where producers can voice their interests and express their needs. Village cotton producers group (GVPC) members are satisfied with union services if the cotton sold and the rebates are paid on time, as well as being adequately and timely informed on the results of the negotiations with market partners, which in turn allow them to plan their cotton production. According to members several factors hamper unions in playing these roles fully and effectively.

In general the village group representatives refer any problems encountered or issues to be resolved to the UCPC. However, when only a minority within the village group has a particular problem and the person representing the group is not part of that minority, it is unlikely that the problem will ever reach UCPC level. Even if the problem or issue is communicated, the majority of board members have to agree before submitting them to the next higher levels. The Boukoumbé UCPC therefore applies a 'subsidiarity principle': problems that can be solved by a GVPC should be solved at this level. This measure is also prompted by the financial situation of this union, which does not allow for frequent travel. Sometimes cotton producers communicate their problems to the UCPC technical staff in the field rather than through their group's representatives, because they feel the staff will communicate their problems to the UCPC more effectively than their board members. In the same vein, agricultural extension agents can, and do, also communicate producers' problems to the union.

Mechanisms exist through which the executive board members report back to the other UCPC board members. However, this process does not always function properly and other board members often feel reluctant to exercise their rights. A key factor is the financial resource base of the UCPC, which largely depends on the volume of cotton produced in the district. It determines the possibilities for representing the union in meetings at other levels (e.g., through daily allowances for board members while travelling) or visiting member village groups (e.g., availability of transport for technical staff). This seems to be one of the main reasons why members are frequently not well-informed about the cotton market, the effectiveness of chain mechanisms and the availability and/or quality of contracted services with third parties, and therefore fail to exercise well-argued influence on decision-making concerning these issues.

From the interviews conducted, it becomes clear that producers at grass-roots level and village group members do not always feel well-represented by the UCPC. They perceive that board members are increasingly becoming part of closed networks and are just looking for their own 'piece of the pie'. Furthermore, interference of local politics with unions also seems to be increasing, with some board members linking up with the political powers.

SWOT analysis

Table 2 summarizes the strengths, weaknesses, opportunities and threats that the three unions face; these have been identified during interviews and through a review of literature and documents. Interviewees and other sources emphasize the overall governance crisis both within the sector, where the AIC and other actors have difficulties enforcing contracts and rules, and in cotton producers organizations, which are increasingly becoming objects of rent-seeking.

Table 4: UCPCs SWOT analysis for social inclusion

Strengths	Weaknesses
All cotton producers can become members. Broad membership-base covering all villages.[K] Clearly defined and disseminated statutes and by-laws for elections. Consolidated mechanisms for funding of organizational functioning. Equipment and infrastructure.[K D] Well-trained technical staff.[K D] Database on cotton production (cotton statistics).	Focus only on cotton production and less attention for agricultural diversification. Illiteracy among members; especially women. No member registration and up-to-date records. Institutionalization of informal eligibility criteria. Lack of transparency in resource allocation (e.g., embezzlement of funds).[D] Failing management of collective repayment obligation. Little understanding of the relationship with the district extension service (funding through cotton levies). Weak communication flows between unions and village groups.
Opportunities	**Threats**
Reforms foreseeing an increasing role for producer organizations. National policy of agricultural diversification (diversification-oriented projects). Guaranteed floor/basic prices for cotton. Member of a national network of unions. Related to a network of rural banks.[K] Being solicited for contributions to local development.	Declining cotton prices on the world market. Contract clauses not respected by partners (e.g., delay in input procurement and cotton payments). Lack of quality control of inputs. Declining yields and quality of cotton.[D] District collective repayment obligation, which has been imposed. Multiplicity of 'cotton taxes'. Politicization. Land tenure (pressure on land).[B]

B = Boukoumbé; D = Djidja; and K = Kandi; factors that are particularly relevant for these UCPCs.
Sources: Tama-Imorou (2006) and SNV Conseils (2005 and 2006).

Role of UCPC in access to services

UCPCs provide a variety of services to their member organizations (GVPCs) and to their individual members:
- Management support for input supply (e.g., centralizing orders for inputs), marketing of cotton (e.g., planning transport of cotton to ginning mills), and payment of fees and rebates.
- Training of board members of village groups on handling of management documents, as well as management of cooperative groups (e.g., information on rules and regulations in force).

- Information and training of cotton producers on new production technologies for pest management in collaboration with the district extension service (CeRPA) and private input suppliers.
- Support to village groups when registering with local authorities (e.g., adapting statutes and by-laws).
- Financial assistance when village groups lose cotton through bush fires or during transport.
- Mediation in conflicts between cotton producers and livestock holders when roaming cattle destroy crops.

Members cite the three first services as the core business of a UCPC. Cotton producers consider information and training on the use of new inputs as crucial for the quality of cotton fibre, since prices are fixed according to grading classes. As a FUPRO staff member declared: 'Our producer organizations work for all members, without any distinction' (Kamminga, 2005). In other words, everybody producing cotton, whether he or she is a small or large farmer, has equal access to the union's training services. However, the way training needs are communicated to the union level (see above), and the manner in which trainees are selected, influence who will benefit from training. Training is being organized through a 'cascade' or 'training of trainers' (ToT) approach: trainees are selected from village cotton producers groups and these trainees in return have to train other group members. In practice, board members from village groups (often the president and secretary) are selected without considering their individual training and communication skills.

Each union has also appointed an information officer *(chargé de vulgarisation)* among its board members who is responsible for coordinating training sessions that are organized by third parties, such as the district agricultural extension service. Djidja UCPC reinforced its information officers' role as liaison officer by appointing three information officers for each of its three sections instead of one for the entire union. This allowed for better organization of information flows and member training.

Training services provided by third parties have changed considerably by introducing more demand-driven approaches and enhanced involvement of the private sector. The district extension agents identify information and training needs at the village level in a participatory way. They may even distinguish several sub-groups, such as women and young farmers. A synthesis of village needs is made at the district level and may lead to a joint training programme with the UCPC that is implemented on a cost-sharing basis, as in Kandi. Over the last few years, private input supply firms have also started funding and organizing training sessions on cotton crop protection techniques; notably in districts where they handle a substantial share of the input market (Sogbohossou *et al.*, 2005; Dotia *et al.*, 2006).

Since 2000, the district extension service operates under a framework agreement between the Ministry of Agriculture and AIC, whereby extension is partially funded through cotton levies (e.g., for recruiting extra field officers and by supporting operational costs). The agreement defines outcome and impact indicators such as: number of village group representatives that have been trained in participatory needs assessment; adoption rates of improved cotton production techniques; and increases in cotton yields. The national extension service even provided UCPCs with a list of criteria to assess the performance of extension agents. This offers cotton producer organizations a tool with which to monitor the quality and impact of extension services. However, few unions and even fewer village groups are informed about these mechanisms, since decisions are taken at AIC level. It is also worthwhile noting that efforts to reach the agreed outcome and impact targets have made extension services pay more attention to the more important cotton producing areas and the larger cotton holdings. (For a more detailed description, see also Sogbohossou *et al.*, 2005.)

Notwithstanding the intention of the 2005 reforms for the unions to focus exclusively on cotton production and marketing, the UCPC participation in local development has been maintained (see Box 3). The UCPCs continue to provide services that reach beyond their membership base. Members complain that the allocation of funds for these purposes lacks transparency and contains risks of politicizing the union. Financial contributions from the union are more often the result of social pressures than properly planned and well-argued investments; as some said: 'When you say no because it hasn't been budgeted, they say they will destroy you. They will set up the community against us'. Local authorities increasingly establish 'cotton taxes' to mobilize resources for the municipality (Baltissen and Hilhorst, 2005: p. 39).

Box 3: Participation in 'local development' by UCPCs

Contribution to funding community infrastructure such as schools, health centres and rural radio stations (Kandi and Djidja);
Financial assistance by local authorities (Kandi, Boukoumbé and Djidja), in particular the municipal authorities (Mairie), and civil society organizations (Kandi);
Scholarships for promising pupils of local high schools (Kandi and Djidja) and students at the national university (Kandi);
Sponsoring of local football teams (Kandi);
Financial assistance to persons facing problems such as illness or death; beneficiaries are usually local opinion leaders (Kandi).

Concluding remarks

The ongoing liberalization and privatization of the Benin cotton sector has led to a rather complex and elaborate system that requires close cooperation between stakeholders and an effective enforcement of rules and contracts (see Table 1).

Difficulties when implementing reforms, in combination with declining world markets prices, have led to violations of the rules at the expense of cotton producers. In addition, producers' organizations, such as the UCPCs, have a long way to go in their intended transformation: from standard supply organizations under the supervision of state services, to more tailor-made, member-led organizations that represent and defend cotton producers' interests vis-a-vis cotton chain operators, service providers and district authorities.

The latest UCPC reform included a quasi-unique focusing on cotton, which had a narrowing effect on the membership basis of the organization. This probably has more consequences for women than men, because women agricultural processing groups could be member of the old unions; these now seem to be excluded. Even when they are UCPC members, women face other barriers in exercising influence, such as land tenure, traditional culture and their 'perceived' lack of capacities. These factors explain the rather small impact of the informal women quota system for union boards, which is promoted by FUPRO but without adequate capacity strengthening activities for women. Skill development, ranging from literacy training to managing groups, seems to be an appropriate way for women to gain recognition and become leaders.

Crop (cotton) and gender (women) are two factors that aggravate social exclusion within UCPCs. Even when it has been officially declared that all cotton producers have an equal opportunity to become board members and should receive equal access to services, the reality is quite different. Although diversity and up-to-date membership records are not an issue for union membership, the volume of cotton produced, as an indicator for financial assets, is a well-known and determining factor. The formal rule of 'one member group, one vote' is gradually being abandoned to the benefit of weighing the production capacity of a group when electing representatives. Other excluding factors are: belonging to a minority group whose language is not spoken during union meetings or who have insecure land rights; and living at the periphery of district capitals that are increasingly becoming the decision-making centres for local development. The result is a rather closed network of union leaders who increasingly adopt rent-seeking behaviour. Despite detailed statutes and by-laws, UCPCs lack effective countervailing mechanisms against exclusion on which members can rely. The overall effect is that ordinary members, who are less wealthy, risk feeling excluded and this threatens the entire social cohesion of unions.

Governance problems within the cotton sector have put the bonding and bridging social capital of the UCPC member organizations (GVPCs) under enormous strain: farmers abandon cotton growing because of delays in payment and 'break-away' producer networks appear. However, linking social capital remains a major asset: unions cover all districts of Benin, which are the decentralized administrative entities par excellence, and are part of a national federation (FUPRO). UCPCs also continue to play a key role in district development through direct funding of community services, which underscores the potential role of cash crops in supporting local economic development, and indirect funding (through cotton levies) of agricultural services. The contribution to district development may have an inclusive effect beyond the reach of mere cotton growers but is threatened by declining cotton prices, risks of politicization, and failing budgeting and accountability mechanisms within UCPCs. Funding of agricultural services by the AIC, of which cotton producers are members, through performance contracts with service providers that are held accountable, also offer opportunities to make services more inclusive. This is particularly important, since agricultural diversification is a national policy priority (NCDFP, 2002: pp. 23-28). Basic conditions to enhance the role of UCPCs in diversifying agriculture for livelihoods include: membership diversity is known and taken into account by unions; investments are well-argued, transparent and accounted for by the union's board; and decision-making of funding of services is delegated to district level.

References

Anon., 1995. *Statuts et Règlements Intérieurs Types des Groupements Villageois de Producteurs (GV) et des Unions Sous-Préfectorales de Producteurs (USPP)*. Parakou, Benin.

Baltissen, G., and T. Hilhorst (eds), 2005. *Les premiers pas des Communes au Bénin. Enseignements du processus de la décentralisation.* CEDA/KIT, Bulletin 371, KIT. Amsterdam, The Netherlands.

Gotoechan-Hodonou, H., M. Adomou, and B. Wennink, 2005. Competitive funds for zonal research programmes in Benin. Chapter 4.4 in: Heemskerk, W., and B. Wennink (eds). 2005. *Stakeholder-driven funding mechanisms for agricultural innovation. Case studies from Sub-Saharan Africa*, pp. 72-81. Bulletin 373, KIT. Amsterdam, The Netherlands.

Kamminga, E., 2005. *Reconnaissance visit to FUPRO Benin.* 15-18 February 2005. KIT Working Document. KIT Amsterdam, The Netherlands.

Kouton, T., G. Yorou, G. Nouatin, and B. Wennink, 2006. Orienting research and development for cotton production. The national federation of producers' unions (FUPRO) in Benin. In: Wennink, B. and W. Heemskerk (eds), 2006. *Farmers' organizations and agricultural innovation. Case studies from Benin, Rwanda and Tanzania*, pp. 47-54. Bulletin 374, KIT. Amsterdam, The Netherlands.

Minot, N., and L. Daniels, 2002. *Impact of global cotton markets on rural poverty in Benin.* IFPRI, Washington, USA.

Sinzogan, A.A.C., J. Jiggins, S. Voudouhé, D.K. Kossou, E. Totin, and A. Van Huis, 2006. An analysis of the organizational linkages in the cotton industry in Benin. Chapter 3 in: A.A.C. Sinzogan. 2006. *Facilitating learning toward sustainable cotton pest management in Benin: the interactive design of research for development,* pp. 35-59. Tropical Resource Management Paper No. 82, Wageningen University, Wageningen, The Netherlands.

NCDFP, 2002. *Benin poverty reduction strategy paper.* 2003-2005. Translated from French. Republic of Benin, Cotonou, Benin.

SNV Conseils, 2005. *Avant-projet d'élaboration des plans d'action pour le renforcement des capacités des Unions Communales des Producteurs de Coton (UCPC). Rapport du diagnostic de l'UCPC de Kandi.* SNV, Parakou, Benin.

SNV Conseils, 2006. *Avant-projet d'élaboration des plans d'action pour le renforcement des capacités des Unions Communales des Producteurs de Coton (UCPC). Rapport du diagnostic de l'UCPC de Djidja.* SNV, Parakou, Benin.

Sogbohossou, C.A., R. Fassassi, and B. Wennink, 2005. Public and private funding of agricultural extension in Benin. Chapter 5.2 in: Heemskerk, W., and B. Wennink (eds). 2005. *Stakeholder-driven funding mechanisms for agricultural innovation. Case studies from Sub-Saharan Africa,* pp. 72-81. Bulletin 373, KIT. Amsterdam, The Netherlands.

Wennink, B., and O.D. Dotia, 2004. *Le rôle des organisations paysannes dans le système de recherche et de vulgarisation agricoles. Les cas des UCP de Kalalé et Boukoumbé et de l'ACooBéPA de Tchaourou/Ouèssè au Bénin.* Working Document. KIT/FUPRO, Amsterdam/Bohicon, The Netherlands.

World Bank, 2005. Benin. *Diagnostic Trade Integration Study.* Volume 1. Draft. World Bank, Washington, USA.

Notes

1 A 'floor price' *(prix plancher)* is a prefixed basic price that is valid for all cotton producers in Benin.

2 *Union Sous-Préfectorale des Producteurs*; the *Sous-Préfectures* became *Communes* (districts) under the decentralization laws that became effective in 2003.

3 The entire levy collected on cotton for financing producer organizations is divided as follows among the different levels: about 60% for GVPCs, 35% for UCPCs, and 5% for UDPCs and FUPRO.

4 A price to be paid at the farm gate is defined before marketing the product. However, when actual selling on the world market takes places, prices may turn out to be higher. This allows for paying a separate, supplementary amount – the rebate – to producers.

5 The social share *(part sociale)* is a pre-defined amount of money that every member pays when becoming a member. It allows for constituting a fund (i.e., working capital) for the farmer group. This fund can be used to procure equipment or construct infrastructure, which then becomes collective, group property.

IV KILICAFE's role in pro-poor service provision in Tanzania

Adolph Kumburu and Willem Heemskerk

Introduction

Coffee is Tanzania's largest export crop. In 2003, Tanzania produced about 800,000 of the 60-kg bags, or 48,000 Metric Tons (MT), which is approximately 0.7% of the total world output (about 117 million bags). In that year coffee contributed approximately USD 115 million to Tanzania's export earnings. The crop provides employment to some 400,000 families and benefits to 2.12 million family members. Smallholders on average holdings of 1-2 hectares grow about 95% of coffee, intercropped with food crops, and about 5% is grown on estates. An estimated 50% of coffee smallholders use purchased inputs. In 2004, coffee production fell to 38,683 MT, with a value of only USD 50 million, due to climatic variation and lower world market prices. About two-thirds of the coffee grown in Tanzania is mild Arabica (Arusha and Kilimanjaro regions of the north and the Mbeya and Ruvuma regions of the south); Robustas are produced in the Kagera region of the Lake Zone. Mild Arabicas are wet-processed, while Robustas are dry-processed (Parrish *et al.*, 2005; Technoserve, 2006).

Coffee is Tanzania's largest export crop but small-scale farmers have not reaped the potential benefits because of policies that have restricted their direct access to the international coffee market. Unions, traders and farmers are obliged to sell their export coffee through the government-run Moshi Coffee Auction with a minimum transaction of 10,000 kg of green coffee[1], effectively barring small farmers from participating as individuals. Farmers have long complained about the inefficiency of the Moshi Coffee Auction and the fact that it causes unnecessary delays and prevents them from benefiting from the full potential of coffee product sales. The cooperative movement in Tanzania has always played a major role in marketing and price formation, also in coffee. The KNCU (The Kilimanjaro Native Cooperative Union) was one of the first cooperative unions to be registered in 1931 (Heemskerk and Wennink, 2003). Since 1991 coffee unions operate as private entities and membership is no longer compulsory. KNCU presently collects, stores and processes coffee from over 150,000 small-scale farmers, through their village level primary cooperative societies. The Tanzania Coffee Growers Association (TCGA), established in 1945, promotes the interests of large coffee farmers and estate producers. The Tanganyika Farmers Association, which was established in 1955 and currently has some 4,000 small and 1,000 large farm members, also includes coffee farmers (Heemskerk and Wenink, 2003). New smallholder coffee organizations are currently emerging, in response to large overhead costs of the old organizations, resulting in lower prices for the farmers, but also in making use of speciality coffee opportunities in the market. KILICAFE is one such organization, which works specifically for smallholders. KILICAFE is a commodity-based farmers' organization, in contrast to the intermediate network organization MVIWAMO, which is part of MVIWATA.

Coffee research and advisory services have been privatized, partly due to the dissatisfaction of the levy-paying coffee sector with the level of progress through innovation in the sector. Since most coffee producers are smallholders with a great diversity in cropping and farming systems as well as levels of income, special attention is required for socially inclusive service provision in the coffee sector. KILICAFE conducted a study on agricultural service provision in 2005, to identify the constituency and the membership, and the direct or indirect relationships with coffee sector service provision. The study was conducted at three levels: the Farmer Business Group (FBG), Chapter (i.e., intermediate level) and KILICAFE National Management level. A total of 93 individual farmers (FBG members) were interviewed, as well as 12 FBG management committee members. Primary data was collected through interviews using a similar questionnaire for all three levels. Secondary data was collected from various head office documents, previous case studies and the organization's website (www.kilicafe.com), as well as general information from the management. A SWOT analysis was conducted using all the existing information, which was validated in a workshop at national level.

Presenting KILICAFE

Trade liberalization in Tanzania opened the doors for agricultural groups to participate fully in marketing the crops that they produce. 'Technoserve' (TNS) Tanzania, an American-based NGO, started supporting coffee producing groups in 1999. TNS promoted and encouraged independent Farmer Business Groups (FBGs) and assisted them in organizational structure improvement; this in turn led to the formation in 2001 of an apex association: 'the Association of Kilimanjaro Specialty Coffee Growers', later known as KILICAFE[2]. Apart from governance problems in the 'old' farmers' organizations (see earlier) KILICAFE is also a response by farmers to the low coffee prices and the desire to provide quality coffee through a value chain with a stronger role of the farmers' organization in the chain through value-adding activities.

Unions pay a uniform price regardless of the quality of the coffee. KILICAFE's policy is to reward quality by paying differentials in order to create competition to raise the quality of coffee from its member farmers. KILICAFE aims to improve the incomes of its members by earning higher coffee prices, and hence improve their living standards through product quality improvements e.g., use of better processing methods (central pulping units) and aggressive marketing strategies e.g., through direct exports and branding. KILICAFE's sustainability strategy is based on internal 'company standards' and international certification schemes, which both observe:
1. enhanced business practices;
2. processing and production practices;
3. attention for environmental care; and
4. social responsibility, for which it scores 87%, 48%, 54% and 72% respectively (Kumburu, 2006).

KILICAFE is a membership-based organization registered as a Limited Company under the Company Ordinance (Cap. 212), composed of three intermediate-level networks, chapters or KILICAFE branches of registered FBGs. All FBGs are independent organizations formed by smallholder coffee growers and chaired by elected committees. Chapters do not exist as independent legal entities, but operate under KILICAFE as branch offices, with administrative committees formed by members elected from among FBG leaders. A chapter is formed once it consists of atleast ten FBGs and has a potential to produce a minimum of a 100 MTs of parchment coffee in a season. Most of the FBGs were formed after the demise (due to a loss of trust) of the primary cooperative societies. FBGs can have two types of legal status: either being registered as cooperatives under the Trustees Incorporation Act (Cap. 318) or as 'partnerships or associations' under the Business Names Registration Ordinance (Cap. 213) under the Company Ordinance. Of the groups surveyed, all northern groups hold registration certificates under Cap. 318, and all southern groups have registrations from Cap. 213. No actual differences are observed as a result of the two types of registration, apart from the fact that those registered under Cap. 318 have a broader area of operation in terms of mandated activities than partnerships, which are limited to the few activities stated in the registration form. KILICAFE mainly operates in the Arabica-growing areas of the country through the three referred chapters. The distribution of average coffee production per chapter (over the past three years) is as follows: 40% Mbeya, 40% Mbinga and 20% North. The low volumes for the north are attributed to lower acreage under coffee and old coffee trees; 50% of the trees are over 50 years old, according to the Tanzania Coffee Board (TCB) census of 2001 (Kumburu, 2006; Technoserve, 2006). KILICAFE purchased 2,100 tons of coffee in 2005 and had a turnover of USD 3 million in sales at the local auctions and 'Direct Export', which amounts to around 5% of the total national coffee production (See Table 1). Specialty coffee marketed through the national auction yielded a 65% premium, compared to the lower quality blended coffee over the past two years (Parrish *et al.*, 2005; TNS, 2006). Direct exports yielded 150% price premiums for smallholder suppliers (coffee processed in central pulping units, or CPUs). The KILICAFE coffee turnover is expected to reach USD 4.5 million within the next three years.

Table 1: Coffee production from each KILICAFE chapter in kg of green coffee

Chapter	2004	2005	2005/06-Est.
North	255,561	456,845	271,858
Mbinga	593,600	796,589	641,979
Mbeya	288,289	809,305	356,017
Total	1,107,450	2,052,739	1,269,854[3]

Source: TNS, 2006.

The KILICAFE organization finances its activities through:
1. coffee sales (3% of the gross sales is used to run the association);
2. the FBG membership fee is the equivalent of USD 25;
3. bank loans for advance payments and marketing expenses;
4. donations and grants from international organizations and business partners; and,
5. a Fair Trade sales premium.

The activities at FBG level are financed by fixed fees, which are charged from the sales proceeds of the FBGs; these range from Tshs 10-20 per kg of parchment coffee. The amount is fixed at the FBG Annual General Meeting (AGM). KILICAFE also raises funds from donors by submitting proposals for specific activities that focus on benefiting the producer community and members through solving specific problems. As an umbrella organization KILICAFE possesses equipment and rents offices, while the FBG main assets are the Central Pulping Units[4] (a total of 29 FBGs owned central pulping units in 2005, which rose to 55 in 2006) and coffee storage houses, all acquired through TNS credit facilities.

The KILICAFE membership base

FBGs are the 'first level' organization at the grass-roots level of the KILICAFE structure. FBG membership conditions are shown below. A member must:
1. own a coffee farm;
2. be a local resident and member of the community;
3. agree to be trained in ways to improve the quality of coffee;
4. pay an FBG membership fee ranging between Tshs 2,000 to 5,000, which is fixed by the FBG Annual General Meeting;
5. deliver coffee to the FBG for joint processing and/or marketing;
6. accept the FBG constitution;

7. promote the FBG and KILICAFE as a whole; and,
8. cooperate with fellow members to improve the economic and social welfare of the group.

FBG members are individual coffee farmers who own small farms ranging in size from 0.5-2 acres in the north, 1-7 acres in Mbinga and 1-12 acres in Mbeya. Most members of FBGs are producers from the same village or ward, who deliver their produce to the group for marketing. FBG members have membership cards and are registered in the FBG ledger; they are different from non-member farmers in the common ownership of FBG assets such as coffee stores, processing equipment at the CPU, and other infrastructure such as a clean water supply system, etc. A member may be disqualified if he/she does not deliver any coffee during the season, but no minimum amount of coffee to be delivered has been established. The only essential requirement is that the individual has a coffee plot and that he/she intends to improve his/her coffee quality. Even very small coffee producers have become members of KILICAFE FBGs. The number of farmer members in FBGs ranges from 25-250, depending on individual groups and demographical conditions in the area, though groups in the north are usually largest. KILICAFE membership is open, and in the recent past has been growing rapidly, both in terms of groups and in the number of members per group (see Table 2). The expansion from the 10 founding groups in 2001 to the present 102 FBGs in 2006 amounts to a total of over 10,000 smallholder members. However, the KILICAFE Board is concerned that uncontrolled growth in membership may affect the quality of services rendered; hence the 2006 AGM issued a temporary freeze on FBG establishment until it can be certain that the association can accommodate more members with quality services. No evident social distinction is identifiable in the KILICAFE FBG memberships.

Table 2: The increase in the number of KILICAFE FBGs over the years

2001	2002	2003	2004	2005	2006
11	37	47	75	93	102

Source: TNS, 2006.

Gender dimensions

The KILICAFE constitution clearly states that member FBGs must state in their respective constitutions that they do not discriminate on the basis of gender, age, religion, race or tribe, as a condition for joining KILICAFE. However, a gender imbalance does exist in FBG membership; although it can be argued that households are members of the FBGs, it is mostly men that are actually registered as members and less than 10% are women, most of these are from female-headed households. Fewer than 10% of FBGs have women on the committees. Out of the 12 FBGs surveyed, only Makisomila in Mbinga Chapter had a woman on the management committee. Traditionally it is the men who are head of the household, own the coffee trees and control the earnings from this important cash crop, but women contribute a significant portion of the required labour, particularly for picking and drying the coffee beans (TNS, 2006). Although the KILICAFE Board is aware of the need to be gender-sensitive, there is as yet no deliberate policy to achieve a reasonable gender balance. However, KILICAFE is convinced that cultural values can be modified and changed over time and is working towards this goal through intensive education efforts. Some changes have started to occur in chapters at several locations, e.g., the chairperson of the North chapter is a lady.

Building social capital

KILICAFE supports the strengthening of FBGs (i.e., 'bonding' social capital, see Box 1), the reinforcement of chapters (i.e., 'bridging' social capital, see Box 2), as well as enhancing the capacity to interact with other stakeholders (i.e., 'linking' social capital) (for details on this terminology see Heemskerk and Wennink, 2003). The planning of KILICAFE activities is carried out by management and approved by the Board of Directors. Overall monitoring is vested in the Executive Director, assisted by Heads of Departments as well as Chapter chairpersons.

> **Box 1: Main activities related to the strengthening of bonding social capital supported by KILICAFE**
>
> - Recruiting member farmers by promoting FBG results and services.
> - Maintaining membership register records.
> - Maintaining members' coffee-delivery records, Ledger Cards and Control Sheets.
> - Collecting membership fees and other contributions from members.
> - Collecting coffee from members and delivering coffee to mills.
> - Paying sales proceeds to FBG members according to coffee weight notes.
> - Collecting inputs requirements and distributing inputs to members.
> - Maintaining business records and acounts.

KILICAFE also facilitates links with other stakeholders and, as such, strengthens the linking component of social capital. KILICAFE is a member of the regional organization EAFCA (The Eastern Africa Fine Coffees Association), which is a regional coffee promotion initiative formed by 10 countries, with its head office in Kampala, Uganda (Kumburu, 2006). Nationally KILICAFE is a member of the Tanzania Coffee Association, involving a cross-section of industry stakeholders from the private and public sector, as well as the privatized Tanzania Coffee Research Institute (TaCRI). KILICAFE has a reputable relationship with the Tanzania Coffee Board (TCB), which supervises the coffee sector and issued the KILICAFE's export licence. KILICAFE hopes to convince the TCB Board to reduce or abolish particular charges such as license fees,

> **Box 2: Main activities for strengthening of bridging social capital supported by KILICAFE:**
>
> - New FBG formation and membership growth. Committee members receive applications from aspiring groups and KILICAFE staff visit to evaluate the group before it is accepted as a member.
> - Assistance to KILICAFE office at chapter level on day-to-day running activities, and to ensure that KILICAFE policies are followed by FBGs.
> - Assisting KILICAFE chapters to identify and appoint input suppliers for their FBGs.
> - Distributing payments to FBGs and farmers (advance, interim, and final) remitted from KILICAFE Head Office.
> - Promotion and publicity of KILICAFE activities in the chapter through meetings conducted in FBGs.
> - Facilitating the conduct of Chapter AGMs as well as quarterly chapter committee meetings.
> - Mediating between members' conflicts and problems arising within FBGs.

coffee cess, etc. and hopes to convince the TCB, plus central and local governments, to reduce various taxes on coffee.

KILICAFE is an authorized seller of member FBG coffees at the National Auctions market and in direct exports. KILICAFE has contractual arrangements with TNS for technical assistance and advice on coffee-related activities and business management[5]. Similarly, through a memorandum of understanding with Taylor Winch Tanzania Ltd (TWT), a private coffee trader, KILICAFE receives support in export handling[6]. As an independent association KILICAFE has no formal relationships with traditional authorities or civil society organizations, except in such areas where collaboration provides greater benefit to its members.

Representativeness of KILICAFE

KILICAFE plays an active role in alleviating poverty of its members by increasing their incomes from coffee sales through quality improvement and higher market access. This is demonstrated by the premium prices members receive, as compared to non-member coffee producers in the same growing regions (see Box 3).

FBG Committee members, as well as the KILICAFE Chapter Committees and KILICAFE Board, do not think social exclusion is a threat to the survival of the association. All players at all levels strongly believe that KILICAFE is an equal opportunity organization and is non-discriminatory in all its dealings. The members gave KILICAFE a high rating for its social responsibility (72%) compared to international standards (Kumburu, 2006). The phenomenal growth in membership is a testimony not only to the quality of the services supplied but also to the association's social responsibility towards coffee growers. KILICAFE excludes non-coffee-growing households, while excluding farmers in the Robusta coffee growing areas, as the main focus is on Arabica speciality coffee. Although KILICAFE is not a member of

MVIWATA, efforts are underway to strengthen relationships between the two organizations, while coffee FBGs can already become members of the MVIWATA network.

Role of farmers in KILICAFE

FBGs have individual constitutions endorsed by all members at the FBG Annual General Meetings. These meetings: elect the FBG Management Committee (seven members, including chairperson, secretary and treasury); provide guidance for operations, and set membership fees. At the chapter level the AGM is open to all individual farmer members to attend as observers. The chapter AGM receives, discusses and approves the annual business plans submitted by the chapter committee, which is a management committee of seven members drawn from leaders of the participating FBGs. The chapters' main responsibility is to provide a link between the FBG and the KILICAFE Governing Board, representing the interests of farmers from the respective regions. The highest authority at KILICAFE national level is the AGM, which sets policies to guide the head office and the activities of the chapters'. The AGM appoints the Board of Directors, which is drawn from the chapter committee members and elected every three years at the national AGM. The Board meets quarterly, but the KILICAFE management team, headed by the Executive Director who is appointed by the Board and who is secretary at all board meetings, handles the daily activities. The AGM decides on the costs of the contracted services to be provided by KILICAFE to the FBGs. A member can be elected to any of the decision-making bodies, if he/she is:
1. a coffee farmer;
2. a member of an FBG; and,
3. a leader in one of the FBGs.

Box 3: KILICAFE's record in raising coffee prices. A TNS statement quoted in a press release from TNS in April 2004

'KILICAFE has consistently earned higher prices for its members by providing efficient services, including providing credit and marketing support. In the 2002 and 2003 seasons, the prices KILICAFE members obtained represented over 65% premiums above the average price paid in Mbinga, with some farmers receiving much higher premiums. KILICAFE's performance this past year has been outstanding and I commend KILICAFE's directors and management for their professionalism and hard work,' commented Paul Stewart, Technoserve's Coffee Marketing and Finance Coordinator. Stewart added, 'I am confident that many more smallholder farmers will achieve these high prices in the future, as Tanzania retakes its rightful position as one of the world's leading specialty coffee origins'.

The management and the decision-making bodies in KILICAFE are accountable to KILICAFE members (the FBGs). The latter are informed about the operations and financial affairs of their association through periodic meeting reports, from headquarters to chapter committees and finally to members through the FBG committees. These include coffee sales accounts, which are discussed at FBG meetings before payments are made to individual farmers. The KILICAFE and FBG Constitutions state that government officials and political party leaders cannot be appointed to governing bodies or to the management of the association; this makes it impossible for the political party leaders to hijack the organization. All chapter offices are run by qualified KILICAFE staff who manage chapter operations and finances; two per chapter (a qualified accountant and an operations officer). The total KILICAFE workforce comprises 11 employees. KILICAFE has yet to set aside a budget for training its leaders and staff. Currently the little training KILICAFE staff receive is paid for by the donor organizations such as TNS.

SWOT analysis

KILICAFE analyzed its main strengths and weaknesses through a self-assessment or SWOT analysis; the results have been summarized in Table 3. The main focal point of the analysis is on social inclusion, but not particularly focusing on agricultural innovation.

Role of KILICAFE in access to services and service quality

KILICAFE plays a major role in the development of the linking component of social capital (see chapter 2, figure 3). This includes the interactions with agricultural service providers, such as research and advisory services, financial and input services as well as links to markets, traders and millers. Apart from such facilitation, KILICAFE itself has become a service provider for its members at both FBG level (e.g., pulping services), at chapter level (FBG strengthening) and at national level (e.g., marketing and financial services).

KILICAFE core activities at national level include representing its members at national and international forums, such as policy dialogues with government bodies, coffee industry organizations and participation in marketing exhibitions and conferences (Specialty Coffee Association of America, SCAA). Services provided include credit links and financial management of loans for working capital and CPU establishment. KILICAFE sources finance from donors and/or financial institutions to purchase CPUs and issues repayments from coffee sales to the FBG on 4-year-term loans. Input credits are organized at chapter level, whereby the chapter AGM sets limits on how much to spend on inputs per kilogram. These credits are non-cash loans; they are mere guarantees to input suppliers for future payments. Marketing is

Table 3: KILICAFE SWOT analysis through self-assessment

SWOT	Description
Strengths	Non-discriminatory constitutional policies.
	Leadership is adequately knowledgeable and competent and originates from stakeholders.
	The management is qualified, skilled, kind, honest and motivated.
	KILICAFE is a legal entity registered under Companies Ordinance (Cap..212)
	The Board of Directors is elected on a rotational basis and is accountable to the members.
	Participatory planning activities move from FBG to KILICAFE and vice versa.
	Selling certified specialty coffee processed by CPUs.
	Transparency in all transactions and activities carried out by KILICAFE, thus enabling items to be traced.
	A set code of ethics to account for money and stocks.
	KILICAFE has a good reputation amongst sector stakeholders.
Weaknesses	Inadequate funding for investments.
	Limited number of female members.
	Lack of means of transport at headquarters and branch offices.
	Lack of owned office/residential premises.
	Inadequate communication facilities e.g., telephone, Internet.
Opportunities	Access to international specialty coffee buyers.
	Collaboration with national regulatory bodies.
	Liaison with international institutions.
	Growing market for specialty coffee around the world.
	International recognition status.
	Exposure through internal and international fora and symposia.
Threats	Volatile commodity prices in international markets.
	Changes in government policies for the industry.
	Drought.
	HIV/AIDS and malaria have become major threats because they deplete the labour force.
	Rural/urban migration. The youth migrate from villages to towns leaving FBGs with only elderly residents.
	Lack of financial sustainability, dependence on donor and bank guarantees.

done by sending green coffee samples to the Coffee Board (for buyers at local auctions) and direct posting to overseas coffee roasters (for direct exports). KILICAFE provides technical advisory services and training, such as training farmers on quality production methods, training in CPU operations and business management, provided at FBG level by conducting seminars that are open to all members.

In addition, leadership training is conducted at chapter level to all FBG Management Committee members (chairpersons, secretaries and treasurers)[7], empowering smallholder farmers to own fixed assets, which can be used as collateral for bank loans. KILICAFE also provides communication services such as a quarterly newsletter, radio broadcasts, as well as a website (www.kilicafe.com), all containing information on coffee market price trends, a farm activities calendar, association events and activities, and other new developments.

FBG members also demand other services. Some want KILICAFE to clearly specify coffee processing quality standards and ensure adherence by all FBGs. There is a need to use only recommended technologies to achieve uniformity in quality and thereby premium coffee prices, although low prices also influence the adherence to quality-enhancing standards. FBG members want KILICAFE to make proper arrangements for agro-input supply well in advance. KILICAFE members want the government's and institution's dealings with health matters to increase capacity and invest more in the fight against malaria, HIV/AIDS and other diseases. Malaria, HIV/AIDS and other epidemics in village communities pose a serious threat due to the loss of labour for farm activities at household, community and national levels (Source: this study).

Concluding remarks

One of KILICAFE's key achievements is that (through the FBGs) it has been able to reach the minimum supply threshold of 10,000 kg of green coffee necessary to access the Moshi Auction, as well as obtaining a direct export licence outside the Moshi Auction. The CPUs were an important innovation in attaining these outcomes, and provide a significant contribution to enhanced incomes at smallholder level. KILICAFE's main achievements for its members concern the fact that the prices received by KILICAFE members (over the course of the last three years) have been significantly higher than those reached by comparable coffee marketing groups in their areas of operation. This factor alone will ensure growth in membership, whether in the number of FBGs, individual membership or both.

Complementary services that have been provided by KILICAFE include access to coffee inputs, advance and final payments, training in best farm practices, exchange visits between chapter committee members, facilitation of farmers to attend industry coffee trade exhibitions, etc., giving members added advantages over non-members. KILICAFE maintains regularly audited books of accounts for each FBG transaction, through a centralized computer-based accounting system. The accountability of FBG committees to members, of chapter committees to FBGs, of KILICAFE staff to the Board, and of the Board to the AGM, ensures accountability at all levels of the association, and gives members confidence in good governance.

Monitoring, especially of the accounting and procurement functions, is carried out regularly by the KILICAFE Head Office through scheduled and random visits and checks. KILICAFE, in collaboration with her partner, TNS, has continued to provide technical services such as the introduction of alternative crops to coffee (e.g., the *Artemisia annua* for production of anti-malaria drugs in the north), credit to purchase CPUs and FBG training in business skills. Employment has been boosted in rural communities through CPUs, as each CPU employs a minimum five persons for at least three months to run the CPU on a salary paid by FBG from working capital (often received through a KILICAFE loan). The use of CPUs means that farmers have more time for other economic activities, which were previously tied to processing coffee at home. They can now just pick coffee berries, deliver them to the CPU and get paid an advance based on cherry weight.

Farmers' organizations such as KILICAFE play an important role in supporting the rural Tanzanian economy. Such organizations can also contribute to improving the standards of living for rural families if they are well supported by the public sector in terms of regulations. To some extent farmers' organizations have replaced the roles of the now defunct cooperative unions and their primary societies. KILICAFE will consider itself successful once all its member FBGs are able to improve the quality of their produce through adequate supply of farm inputs and central processing.

From the above analysis it can be concluded that smallholder coffee farmers are well-represented by KILICAFE and that the association's focus on smallholders has resulted in pro-poor technology such as CPUs, which result in premium prices being received. On the other hand, KILICAFE was not found to have a special programme for vulnerable groups in the coffee-producing areas, such as in relation to gender and HIV//AIDS. Farmer members are committed to coffee production and FBG membership. The planning and budgeting cycle is understood by members of the FBGs and is sufficiently transparent. The organizational resources and capacity at FBG level are low, since most FBG Management Committee members have a low level of education and skills, but they do have knowledge of coffee production management, as illustrated by the expressed need for, and use of, farm implements and agro-inputs. However, KILICAFE's development is hampered by insufficient training facilities and lack of budget, further aggravated by poor communication infrastructure in rural areas, which increases the operational costs of the FBGs (including for their own capacity development). Coffee farmers are generally not the poorest of farmers, as even smallholder coffee producers receive a cash income. KILICAFE does not directly represent those farmers in coffee-growing areas that do not grow coffee but have focused instead on food production such as maize or bananas. Concerning new technology, KILICAFE has no special agenda to influence research and extension other than for coffee production[8].

References

Heemskerk, W., and B. Wennink, 2003. *Farmers' organizations in agricultural innovation development. Review and cases from Tanzania and Benin.* Position paper, December 2003.

Kumburu, Adolph, 2006. *Sustainability Strategy. Association of Kilimanjaro Specialty Coffee Growers.* EAFCA 3rd WWC Conference, 18th February 2006. Available online at URL: www.eafca.org/wwc/presentations/WWC/Saturday/Meru/0900-1000/Adolph%20Kumburu_KilicafeSustainability.pdf

Parrish, Bradley, D. Valerie, A. Luzadis, and William R. Bentley, 2005. *What Tanzania's Coffee Farmers Can Teach the World: A Performance-Based Look at the Fair Trade-Free Trade Debate.* Sust.Dev. 13, 177-189(2005). Published online in Wiley InterScience.

Schouten, Chira, and Willem Heemskerk (Eds.), 2004. *Farmers' organizations in agricultural innovation. Case study from Tanzania.* Mviwata, Mviwamo, AKSCG, KIT, DRD.

Steenhuijsen Piters, B. de, W. Heemskerk, and F. van der Pol, 2005. The public and private agricultural research discourse in Sub-Saharan Africa: A case of Romeo and Juliet?. In: Ruben, R. and B. de Steenhuijsen Piters (eds), 2005. *Rural Development in Sub-Saharan Africa. Policy Perspectives for Agriculture, Sustainable Resource Management, and Poverty Reduction,* pp. 43-62. Bulletin 370, KIT. Amsterdam, The Netherlands.

Technoserve, 2006. *Case study on Coffee in Tanzania. Realizing Rights, The Ethical Globalization Initiative.* Available online at URL: www.realizingrights.org/trade/Aid_for_Trade_Tanzania_May06.pdf

Technoserve/USAID, 2006. *Building a competitive Coffee Industry in Tanzania.* Available online at URL: www.africanhunger.org/uploads/articles/4f40ec901cdfc0860654c5faf8816bbf.pdf

Notes

1. Picked cherry coffee is converted into parchment coffee by removing the outer skin (pulp) and drying, while parchment coffee is processed into an exportable green coffee (or clean coffee) through curing (or milling).
2. www.kilicafe.com
3. Expected to reach 2,500 MT.
4. CPUs are where 'cherry coffee' is converted into 'parchment coffee'. The process includes selecting ripe cherry coffee, removing pulp, fermentation, washing and drying.
5. There is a nominal fee of 1% of the annual net sales paid by KILICAFE to TNS for services rendered.
6. TWT provides a fee-based service based on the quantity of coffee directly exported.
7. Outsource experts from academic institutions and/or TNS are used to provide the training sessions.

8   The privatized Tanzanian Coffee Research Institute where KILICAFE is represented on the Board is no longer focusing on coffee farming systems, as was the case in the public research period, but on coffee production as such (Steenhuijsen Piters *et al.*, 2005).

v ACooBéPA's role in pro-poor service provision in Benin

Clarisse Tama-Imorou and Bertus Wennink

Introduction

Cashew trees are grown in central and northern Benin as a cash crop on both large-scale plantations that were installed by state services and are now being leased to private firms for exploitation, and on smallholder farms (Sedjro and Sanni-Agata, 2002: p. 9). Cashew is considered an alternative to cotton, to diversify livelihood systems for small-holders in this area. Because of declining cotton yields in southern and central Benin, frequent delays in payments to producers for their marketed cotton, and growing support for alternative product value chains, cashew production has risen rapidly over the last few years. Also, cashew as a cash crop is less dependent on imported inputs than cotton (Matthess *et al.*, 2005: pp. 48 and 174). As a result of these developments, member organizations from the national federation of agricultural producers' unions (FUPRO), which still mainly depends on cotton revenues to fund its operations, has started organizing members into local village cashew growers groups for cashew production and marketing in the Donga and Atacora regions, northwest of the Ouèssè and Tchaourou districts (see for example UDP Atacora/Domga, 2004: pp. 8 and 12).

Another initiative besides the ones supported by FUPRO has been the establishment of the *Association des Coopératives Béninoises de Planteurs d'Anacardier* (ACooBéPA), which is an association of cashew growers from the Ouèssè and Tchaourou districts in central Benin. Ouèssè and Tchaourou are two districts where the area used to cultivate cashew is increasing at a fast pace, in part because climate and soil conditions are favourable for a variety of crops, especially cashew. Furthermore, relative low land pressure attracts farmers from the northern regions to establish new farms and plant cashew trees. ACooBéPA only covers the aforementioned two districts and it is difficult to localize its headquarters, since the association does not have an office building of its own. ACooBéPA has been created with support from the NGO DEDRAS-ONG. The establishment of an umbrella association was the logical follow-up to organizing cashew growers into cooperative-type village groups, which was facilitated by an earlier project that aimed to develop the cashew value chain.[1] During implementation of this project, DEDRAS-ONG was also the lead organization in organizing cashew growers for supplying cashew to markets. Another, second project that aimed to support the development of the cashew chain in turn contracted DEDRAS-ONG to support these village groups and particularly to help strengthen the capacity of the newly formed association.[2]

Both development projects referred to above are examples of numerous so-called 'agricultural diversification and marketing' projects that have been implemented in Benin over the last five years. These projects provide support for improving production, processing (new technologies) and marketing of targeted cash crops, and for organizing village producer groups through intermediate organizations, alongside the national agricultural extension service (CeRPA). The national agricultural extension service operates in all districts in Benin and works closely with the existing district producers' unions in Benin that are member of FUPRO. In order to actually provide such support, local NGOs, such as DEDRAS-ONG, are often being contracted through the development projects to provide and manage advisory services staff and other field personnel (Dotia *et al.*, 2006: p. 56). The second support project (PADSE), for cashew growing and marketing, which contracted DEDRAS-ONG, also contracted the national agricultural research institute (INRAB[3]) to provide certified planting material (or seeds) and develop improved planting techniques. Project funding thereby gave a new impulse to innovating cashew growing. Until then this was rather limited to sensitizing and informing smallholders on cashew planting, rather than marketing through the district extension service. Agricultural research produces technical fact sheets that are to be used by DEDRAS-ONG extension agents and trainers.

ACooBéPA is therefore an example of a farmers' association that is being used by a project to improve and target service provision for cashew growers, alongside the 'usual' agricultural extension service. The association's level of inclusiveness and its role in service provision to its members were evaluated on the basis of interviews with individual ACooBéPA members and focus groups in 2005, as well as a review of available documentation.

Presenting ACooBéPA

The main reason for creating ACooBéPA was to reduce the number of intermediaries involved in the cashew supply chain and provide a more direct interface between producers and buyers so that producers, most of them smallholders, can influence price setting. Cashew producers are organized into producer groups at the village level (GPA), which function as cooperative groups and more or less follow the statutes and by-laws of the cotton producers' groups[4]. Most village-level cashew growers' groups are registered (since 2001) as cooperative groups with the extension service. These village groups then form a local union at the sub-district level (ULGPA), which also has a cooperative status, and those from both the Ouèssè and Tchaourou districts are organized under the ACooBéPA. The association seeks to function as a cooperative union.

Members present different reasons for creating ACooBéPA. Some seek the origin in experiences of members abroad, for example in the Ivory Coast and Ghana, where smallholder cashew growers have been successfully organized. Others relate its creation to a group of buyers who expressed their desire to be able to meet with a representative of the producers as an intermediary. This was certainly the case for association developments during the first project (ANFANI) mentioned earlier, involving an enterprise that was funded through a public-private partnership. The enterprise was established as an alternative to the mainly Indian cashew buyers; it facilitated the export of raw cashew nuts and the participation of producers in managing the company through holding shares (Verhagen, 2004: p. 16).[5]

Since the production of sound planting material and appropriate management of cashew plantations were considered to be key factors in providing quality cashew nuts, the second project (PADSE) focused on training of nursery gardeners and planters and on providing certified planting material (Dotia *et al.*, 2006: p. 56). According to the information provided, DEDRAS-ONG was involved in, and paid for, facilitating the organization of producers; the NGO maintained this key position during the second project that provided support to ACooBéPA.

ACooBéPA has an elected board of directors *(bureau)* including: a president, a secretary general, a deputy secretary general, and a treasurer; four appointed managers (communication, training, organization and social affairs); and two special counsellors. ACooBéPA is entirely run by members on a voluntary basis, since the association doesn't have the financial resources to employ staff. However, for practical reasons, ACooBéPA often relies on supporting organizations such as DEDRAS-ONG to train its members (e.g., providing trainers and classrooms) or represent its members at meetings with local authorities and services (e.g., organizing transport facilities for ACooBéPA members).

ACooBéPA pursues several aims that include:
1. supplying improved planting materials and planting techniques to producers;
2. organizing the marketing of cashew nuts at remunerative prices;
3. establishing long-term relationships with buyers;
4. facilitating access to credit for its members;

5. exchanging experiences and information among producers; and,
6. representing members at different levels.

However, according to interviewees, there seems to be no overall consensus among members about the mission and objectives of the association. This impression is further reinforced by the fact that the association's statutes and by-laws have not yet been approved by the local authorities. However, marketing cashew nuts at rewarding prices is one objective with which all association members agree.

The variety of objectives that are being put forward by members are clearly related to and influenced by the projects that support the organization. Outside support is crucial for ACooBéPA: members cite the lack of donors and partners, other than DEDRAS-ONG, among the association's most pertinent problems. ACooBéPA has to survive on 'social shares'[6] and membership fees. Levies on the cashew nuts marketed are another financial source for the association but supply few resources due to the lack of difficulties in organizing collective marketing. This limited level of support poses particular challenges to the association's functioning, since it does not permit it to realize the aspirations of its members in the short term.

The initial lack of financial resources also caused DEDRAS-ONG to finance ACooBéPA's creation and the necessary training. So far, many ACooBéPA meetings have been held at the DEDRAS-ONG offices. Some interviewees doubt whether ACooBéPA still functions and thus really exists, since producers have been increasingly selling their cashew nuts on an individual basis for the past three years. As one cashew grower said: 'We have finished with ACooBéPA for three years now. Our Board is still supported by DEDRAS and we hold our meetings there. However, we, cashew growers from the village groups of Tchaourou, continue to work together and sell our cashew nuts to traders'. Furthermore, ACooBéPA has not yet registered with the district agricultural extension service, which acknowledges and registers agricultural cooperatives on behalf of the Ministry of Agriculture.

Membership of ACooBéPA

Members of ACooBéPA are all smallholders who operate cashew plantations and state their commitment to market raw cashew nuts through the association's member organizations. There are two groups of members: those who grow cashew and own plantations *(planteurs)*, and those who are only involved in collecting and marketing raw cashew nuts *(collecteurs)*. Most women members seem to belong to the second group. In order to be a member of a village group (GPA) one first needs to be a cashew producer, which supposes that one owns (or rents) a plantation. It is not clear how strictly this criterion is being applied when considering the aforementioned presence of 'collectors' and 'marketers' among the membership. Secondly, one needs to pay an entry fee (500 francs CFA[7]) and the 'social share' (2,000 francs CFA). Membership of a local union (ULGPA) is also linked to paying an entry fee (500 francs CFA) and a social share (500 francs CFA) per person. Each member also pays an annual membership fee (1,000 francs CFA) that is divided among the different organizational entities.[8] Village groups are rather large and may have up to 100 members. In some villages all household heads are members of the village cashew growers' group.

At the end of 2004, ACooBéPA had about 550 registered members, with a majority of these coming from the Ouèssè district (see Table 1).

Table 1: Membership of ACooBéPA in 2004

Localities	Individual members (GPA)	Village groups (ULGPA)	Local unions
Ouèssè	546 members of which	19	4
Tchaourou	445 men and 101 women	17	6

Source: Wennink and Dotia (2004).

There are no formal criteria for membership, simply because the agricultural extension service (CeRPA) has not yet accepted and defined the required rules (2002). This may seem surprising, but the district extension service (being the sole local authority to approve such applications by farmer groups), still retains the right to do this, and uses standard statutes and by-laws for cooperative organizations such as ACooBéPA.

Although ACooBéPA does not maintain records on its members (e.g., information about areas under cashew), interviewees feel that the association adequately represents the smallholder cashew growers of the two districts, which distinguish themselves from other smallholders only by having a cashew plantation. It presumably replaces some of their cotton income or is their main source of income. This is the particularly the case of cashew growers that have large cashew plantations, up to 30 hectares, which were part of plantations previously installed by state services.

The majority of the ACooBéPA members belong to the *Nagot*,[9] which is the dominant ethnic group in the region. Other ethnic groups that are member of ACooBéPA include the *Bariba* and the Lokpa, who are mainly immigrants from the northwestern Atacora region, and the *Fulani*, who were originally livestock holders from the more Sahelian regions of Benin and neighbouring countries. The *Nagot* maintain a 'joking relationship'[10] with the *Bariba*. Officially, ACooBéPA strives to integrate and represent all ethnic groups in the area. However, members feel that there is an incompatibility between ACooBéPA's professed goals and the situation on the ground, where non-*Nagot* sometimes feel excluded: 'It is the Nagot who dominate. They don't even speak French in the association's meeting. If you don't speak Nagot, you don't understand a thing'. An explanation brought forward by the association's members for the dominance of the *Nagot*, is that district services and other support organizations also mainly consist of members from the *Nagot* community.

Gender dimensions

Producing cashew basically means having land, which is a family household asset, and planting cashew trees. In this part of Benin, heritage of family-owned land is traditionally organized along the lines of male descent. However, in *Nagot* society women can 'possess' land through heritage or buying. Moreover, because of growing land pressure and the overall monetarization of rural economies, women are now increasingly become landowners, by buying land with revenues from their own income-generating activities (see also Sohinto, 2001). According to ACooBéPA members, women usually have small cashew plantations, less than one hectare, but there are some who have two or three hectares.

The ACooBéPA membership criterion of 'producing cashew' therefore means that both women 'land borrowers' and landowners can be members.[11] According to membership records (see Table 1), about 20% of the ACooBéPA membership are female cashew producers. However, the majority of female members seem to be collectors and marketers. Membership also makes it formally possible for a woman to become a Member of the Board. Out of the 14 board members, five are women (2005), one of which is deputy secretary general. As observed during group sessions with members of ACooBéPA, women speak out freely in front of men during the meetings. Female members of the board explained that their participation goes beyond 'window dressing' and that their participation is neither donor-driven nor the result of a particular ACooBéPA strategy. They state that producing cashew is now considered to be a viable, alternative source of income, which gives you social esteem. Increasing their financial autonomy through marketing cashew nuts also means they are perceived as respected entrepreneurs. Once you own land and have planted cashew trees on it, you can gain and maintain your position. Other assets for a woman, besides land, are skills such as reading and writing, which also help women gain a position on the Board.

Building social capital

The village groups (GPA) and local unions (ULGPA) of cashew growers are community-based groups and are well-anchored in the rural communities. According to the members, the strength of the relationships between village groups and local unions as the building blocks of ACooBéPA depends primarily on the soundness of the financial status of village groups. This is because the association largely depends on the financial resources generated by the village groups as the basis for its functioning and the provision of services to its members. However, firm cooperative functioning remains problematic for two main reasons, at least according to interviewees. Firstly, 'cooperative discipline' among members remains weak, because they sell the harvest as soon as possible in order to gain access to cash. Negotiating large-quantity contracts and keeping collective stocks of cashew nuts should result in better prices for cashew growers. However, members feel that credit facilities in the area (also for pre-funding the clearing of cashew tree plantations to protect against bush fires) are not well adapted to the 'business' of producing and marketing cashew and that they lack appropriate stocking facilities. Some cashew growers continue to receive pre-funding from small private traders, in turn for selling them their cashew harvest. This is considered by many interviewees as 'private sector exploitation'. Secondly, ACooBéPA does not manage to establish marketing contracts with buyers that allow the sale of sizable quantities of raw nuts.

In order to reinforce marketing of cashew nuts on a more cooperative-like basis, ACooBéPA has taken the initiative to reorganize the initial, large village groups into smaller entities (*'cellules'* of about 10 persons). They are expected to be formed on the basis of 'affinity' and foster 'trust', and thereby facilitate collective marketing of raw cashew nuts. ACooBéPA also decided that each village group member should market 100-150 kg of cashew nuts through the group's mechanism and provide 25 kg that are to be sold separately to fund the functioning of the association (Wennink and Dotia, 2004).

Relationships that were established between ACooBéPA (as a producer representative) and private sector buyers of raw cashew nuts during the first-phase project have almost ceased to exist. The association now seeks new buyers outside the 'project setting' with whom it can develop relationships on a more sustainable basis. Therefore the association recently undertook a new initiative aimed at getting around NGOs and other such intermediaries and relying on more traditional networks: community members, whether they are cashew growers or not, who are well-known for their negotiation skills and/or have relatives among buyers or brokers (*courtiers*) were recently solicited by ACooBéPA to negotiate on behalf of its members.

At one time, 'rebates' (*ristournes*)[12] – inspired by the manner in which cotton marketing is organized – were also seen as a possible financial source to be paid by DEDRAS-ONG. According to interviewees, this idea never worked out because there was no clarity about how this would be organized and who would be responsible: ACooBéPA, private traders, the 'project' or DEDRAS-ONG. It is important to note that as a result of all this, ACooBéPA has developed an almost dependency-like relationship with projects and related organizations, for chain operations as well as support services. However, as ACooBéPA's own initiative for negotiating contracts with buyers indicates, the association is exploring new ways to develop relationships with chain stakeholders.

Contrary to the cotton producers' organizations, which are relatively wealthy, ACooBéPA member organizations are rarely approached by village communities when mobilizing resources for village infrastructure. However, ACooBéPA may be contacted by local authorities when soliciting specific support from DEDRAS-ONG, which intermediates with specific donors to obtain funding for village infrastructure, such as community health centres.[13]

Representativeness of ACooBéPA

For the past three years ACooBéPA has faced serious problems due to decreasing market prices (see also: African Cashew Alliance, 2006) and a lack of market outlets. Individual producers or local groups find their own outlets to sell cashew nuts with the aim of receiving ready cash. Until now ACooBéPA has had enormous difficulties in pre-funding labour-intensive operations involved in the maintenance and operation of cashew tree plantations, which seriously hampers the association's abilities to perform as a cooperative that stores and sells raw cashew nuts on a collective basis. ACooBéPA also seems to continue to count on 'projects' in the hope of receiving external financial and technical support to access markets. Lack of sufficient financial autonomy, as well as ('farmer-led') ownership, is the main drive behind this strategy.

Both the individual cashew farmers' search for markets and ACooBéPA's ties with DEDRAS-ONG have consequences for the representativeness and strength of the organization. Collective decision-making is increasingly being broken up and is reverting back into the hands of the district-based cashew growers unions (i.e., bridging social capital), which seems to be enhanced by the way DEDRAS-ONG agents work with producer groups. Although ACooBéPA is a community-based farmers' organization with a majority of smallholders among its membership, the cashew growers' organizations have to 'compete' with the omnipresent village cotton producers' groups (GV[14]) that were created by the extension services. These groups are by far the most important community-based organizations in the area. They organize logistics at the village level for input supply and marketing of products, mainly cotton, and in turn receive cotton levies and discounts to finance their operations. In addition, cotton producers groups are unified in apex organizations at the district, provincial and national levels, which makes them important players at all levels. Both these apex organizations and the district extension service also support the cotton producers in managing their groups.

Roles of farmers in ACooBéPA

According to the interviewees, in order to become a member of ACooBéPA's Board, candidates must be:
1. a cashew producer;
2. a member of the board of a village group (GPA) and of a local union (ULGPA);
3. a designated representative (one of two) for the sub-district; and,
4. be of 'good morality'.

An informal criterion also states that candidates need to be a member of a village group that generates 'sizeable revenue', which has not been quantified. The president also has to comply with informal criteria, such as being devoted to his/her work, being a good speaker and a good negotiator. Additional advantages when it comes to being elected are good knowledge of the French language and being a spokesman for the dominant ethnic group in the region, the *Nagot*. The dominance of *Nagot* on the ACooBéPA Board (9 out of 14 members) makes *Nagot* the language used when the Board meets, and makes others feel excluded. In addition, the ethnic groups (*Bariba, Lokpa* and *Fulani*) claiming to be at risk of being excluded are well-represented in only one district, namely Tchaourou. This may also explain the increasingly independent functioning of district entities.

These criteria were mentioned by members but have not yet been formalized into by-laws and written statutes. Members do not currently perceive this as a weakness, since it allows for 'elasticity' (or flexibility) when organizing elections. However, these (non-existent) rules are sometimes used to explain certain practices (without written evidence). ACooBéPA tries to counteract discontent among members about election results by applying a rotation mechanism that allows all villages and sub-districts (and ethnic groups) to be represented on the Board in turn.

Lack of funds is the main reason forwarded for not organizing the necessary statutory meetings of ACooBéPA since the selection of the current Board of Directors in 2002. Extraordinary meetings are sometimes organized to discuss urgent problems during which board members are designated to solicit support with other organizations. Village groups therefore increasingly turn to DEDRAS-ONG agents to request training and other support. In addition, the village groups continue to negotiate with buyers and sell raw cashew nuts. Interviews with members observed criticism of the apparent strong accountability of the Board to projects and related NGOs, and the lack of downward accountability toward its members.

SWOT analysis

Table 2 summarizes the strengths, weaknesses, opportunities and threats that ACooBéPA faces that were mentioned by members and leaders during the interviews.

Table 2: SWOT analysis for social inclusion

Strengths	Weaknesses
Geographic coverage of two districts with important cashew yields. Well-trained membership, with increasing adoption of improved seed materials and plantation techniques. Small village groups for enhancing trust and affinity.	Lack of official and approved rules that leave few possibilities for members' appeal. Little respect for cooperative principles and weak impact of association on household incomes. Weak collective negotiation capacities. Lack of information circulation at all levels. No storage facilities.
Opportunities	**Threats**
Willingness to organize in an associative form. National policy commitment for diversification of cash crops. Prices are related to quality of cashew nuts (grading and quality classes). Existence of support services to achieve legal recognition and adopt statuary texts. Increased interest in the transformation of cashew.	Facing buyers who are better informed about markets. Lack of appropriate credit facilities (no pre-funding of stocks). Weak autonomy of the association whose relationships are developed and managed through an NGO. Lack of management training service.

The central problem faced by ACooBéPA is its malfunctioning as a cooperative organization that markets raw cashew nuts on behalf of its members who share in the dividends. This is still the main incentive for farmers to become members, but until now a positive impact on farmers' incomes has probably been lacking. Therefore members are no longer committed and look for other opportunities to sell or even quit. Sustainable relationships with other chain operators (i.e., buyers) and appropriate pre-financing modalities (i.e., credits) are key factors for the association's success that are not adequately organized by ACooBéPA and its partners. Developing these facilities is still seen as the sole responsibility of donors and accessing donor-financed project support becomes a struggle between interest groups, with the risks of excluding some of the members of the village communities.

Role of ACooBéPA in access to services

Services directly provided by ACooBéPA to its members are limited to contacting buyers for cashew nuts, negotiating prices and contracting, and informing cashew growers on the periods when bush burning is officially allowed by the local authorities. Bush fires devastate cashew tree plantations and are a nightmare for farmers. They protect their plantations through establishing firewalls, which is a rather labour-intensive measure and requires paid labour from outside family households. Since ACooBéPA has few financial resources and is currently restructuring itself, even these direct services (i.e., any services of just bush fire services) are rarely offered anymore.

The relationships with the main agricultural service provider, the district agricultural extension service (CeRPA), are quite limited. The district extension service has a team of technical experts that provides support to field agents working with all village communities of the district, particularly with the cotton producers' groups that were once created by the extension service (Dotia *et al.*, 2006: p. 58). However, CePRA does not always have specialized knowledge on planting cashew trees and managing plantations.

The training and advisory services on managing cashew plantations are organized by DEDRAS-ONG in close collaboration with the agricultural research institute (INRAB). Its field agents work directly with the village groups when assessing training needs, training cashew growers and private tree nursery gardeners. Each agent supports about 10 local village groups (GPA) and trains them in newly developed techniques. The field agents, in consultation with the group's president, select the trainees without any formal criteria (as far as interviewees know). One criterion is explicitly mentioned: being able to disseminate knowledge and information to other group members during farmer training sessions that are organized by the village group. In order to facilitate coordination and organization of training sessions, ACooBéPA appointed a 'training manager' to its Board. However, interviewees state that, too often, agents only report to the NGO's office without involving ACooBéPA. This makes ACooBéPA's members question the services that the association could effectively offer to its members.

Agricultural research is another knowledge service that is being provided to cashew growers. The agricultural diversification support project (PADSE), provides certified planting material (or seeds) and develops improved planting techniques through agricultural research (INRAB). These services are being outsourced (or 'insourced' from the point of view of the public service) through a contract that defines research priorities identified during the preparatory phase (diagnostic survey) of the project.

Annual assessments of services offered by agricultural research, as well as planning services to be provided, is essentially a matter of discussion between the project management unit and the agricultural research institute. ACooBéPA does not always participate in these discussions and negotiations, while cotton producers' organizations are frequently invited to participate in similar (multi-stakeholder) meetings for assessing research results and identifying research priorities, even for crops other than cotton (Dotia *et al.*, 2006: pp. 58-61). Inviting ACooBéPA members to meetings on research priority setting is often done at the sole initiative of donors or projects. Still, ACooBéPA members appreciate services provided by research that allow for innovating and improving cashew production (see Box 1). Cashew growers feel that after years of neglect, researchers are finally paying attention to their crop. Until recently they relied much more on their own innovative capacities and exchange of experiences, communicated mostly during the association's meetings (Wennink and Dotia, 2004).

> **Box 1: ACooBéPA members explain the impact of newly developed technologies because of:**
>
> - the multi-year funding of the research programme;
> - the clear focus (techniques) and target (smallholder cashew growers);
> - the output of research through performance contracts;
> - the relay function of village group members in training and information;
> - the use of appropriate information supports, such as drawings and radio programmes in the local language.

Working relationships between ACooBéPA's entities (GPA and ULGPA) and the agricultural extension service (CeRPA) are therefore rather limited. The extension service considers that cashew growers' organizations already 'benefit' from privileged support provided by DEDRAS-ONG. CeRPA realizes that the NGO is being paid for its services by a donor-funded project, while extension services are in a more or less permanent financial crisis due to lack of government funds (Sogbohossou *et al.*, 2006).

Concluding remarks

Although member opinions differ about the origin of ACooBéPA, its membership base is firmly anchored at the village level through the cooperative producer groups. As a rather young and small organization, ACooBéPA still has a relatively inclusive character. Women's representation among its membership, and their participation in governing the association, illustrate the inclusive character.

Another important feature of ACooBéPA is its close links with donor-funded projects; this strongly shapes the association's relationships with partners in both the cashew chain and service networks. These projects derived their objectives from an overall analysis of the agricultural sector and the cashew chain in particular, and seem to consider the association as a means to achieve development goals. This had two major consequences for ACooBéPA as a member-based and member-led organization:
1. project-supported activities were not always primarily aimed at institutional development (e.g., linking with credit facilities and contracting raw cashew nut

buyers), and organizational strengthening (e.g., facilitating official approval of statutes and by-laws and training in management capacities); and,
2. the association was more or less 'shielded' from relating with other stakeholders within the chain and sector, which hampered 'learning-by-doing' and building social capital.

As a cooperative union ACooBéPA therefore lacks a sound financial basis, which threatens its existence; the association also plays a relatively small role in service provision to its own members. Interviewees even questioned the *raison d'être* of the association. In addition, the lack of funds limits member meetings and affects the flow of information on new technologies and markets to members who are often illiterate. More importantly, it deprives them of information about markets and the skills needed to increase their market share. The demand for financial means has also resulted in some local, farmer-led initiatives, which put the village producer groups at the centre and rely on existing social capital (e.g., by engaging local brokers to negotiate contracts with buyers). However, at the central level of the organization, the quest for access to financial support, which is almost exclusively provided by an ongoing project and an NGO, has an excluding effect on members who seek board membership but do not belong to the dominant village groups. The exclusion of certain groups from governance structures may in the future further weaken the inclusive character of services. While some ACooBéPA members perceive exclusion in the context of 'ethnicity', the underlying reasons are clearly related to proper functioning and performance of the cooperative. Neither ACooBéPA nor its partner organizations have formal documents (i.e., the association's statutes and by-laws) or monitoring mechanisms that provide a starting point for discussing greater levels of inclusion.

To end on a more positive note: ACooBéPA members appreciate research and extension services provided that allow for innovating and improving cashew production. Impact is felt in better yields and is due to the clear focus and targeting of research plus the use of training and disseminating capacities within the network of village cashew groups. Enhancing the involvement in providing knowledge services (e.g., setting priorities and assessing research results) could increase the impact of these services, since the quality of cashew nuts is the determining factor for price setting and an acknowledged competitive asset of the Benin cashew sector (Assouma and Koloboe, 1999: Section 4.1.3.1).

References

AFD, 2005. *Mission EVA. Evaluation rétrospective.* Available at URL: www.afd.fr/jahia/webdav/site/myjahiasite/users/administrateur/public/plaquettes/EVA-2003-2004.pdf

African Cashew Alliance, 2006. *African Cashew Com. An online resource for the African cashew industry.* Available at URL: http://africancashew.com/countries/facts.php?id=3

Assouma, A.B., and A. Koloboe, 1999. *La filière anacarde dans l'Atacora et la Dongha.* CARDER Atacora/ONS, Natitingou/Cotonou. Available at URL: www.runetwork.de/html/en/articles/document.html?Action=displayDocument&id=1379

Dotia, O.D., T. Kouton, and B. Wennink, 2006. Cooperating with agricultural service providers for research and development. Three district producers' unions in Northern and Central Benin. Chapter 5 in: Wennink, B. and W. Heemskerk, 2006. *Farmers' organizations and agricultural innovation. Case studies from Benin, Rwanda and Tanzania,* pp. 55-62. Bulletin 374. KIT Development Policy and Practice, Amsterdam, The Netherlands.

Matthess, A., E. Van den Akker, D. Chougourou, and S. Midingoyi, 2005. *Le cotton au Bénin. Compétivité et durabilité de cinq systèmes culturaux cotonniers dans le cadre de la filière.* BMZ-GTZ/MAEP, Cotonou.

Sedjro, M.A, and L. Sanni-Agata, 2002. *Analyse du secteur de l'anacarde. Situation actuelle et perspective de développement.* CNUCED/OMC/ICC, Cotonou.

Sogbohoussou, C.A., R. Fassassi, and B. Wennink, 2006. Public and private funding of agricultural extension in Benin. Chapter 5.1 in: Heemskerk, W. and B. Wennink, 2006. *Stakeholder-driven funding mechanisms for agricultural innovation. Case studies from Sub-Saharan Africa,* pp. 83-97. Bulletin 373. KIT Development Policy and Practice, Amsterdam, The Netherlands.

Sohinto, D., 2001. Questions du genre liées aux conflits fonciers: impact sur la production durable des vivriers au Sud-Bénin. In: *Afrique et développement,* Vol. XXVI, Nos 3 and 4, 2001, pp. 67-88, Conseil pour le développement de la recherché en sciences sociales en Afrique, Dakar, Senegal.

UDP Atacora/Donga, 2004. *Réflexion paysanne sur un modèle de commercialisation collective d'anacarde au Bénin.* Forum "Accès au marché des produits agricoles". Travaux realises avec l'appui d'Inter-réseaux Développement Rural et le CTA. Available at URL: www.inter-reseaux.org/IMG/doc/Benin_Commercialisation_Anacarde_UDP_IR_CTA_2004-2.doc

Verhagen, H., 2004. *International sustainable chain development. Lessons from the Netherlands, Benin, Bhutan and Costa Rica.* Bulletin 364. KIT Development Policy and Practice, Amsterdam, The Netherlands.

Wennink, B., and O.D. Dotia, 2004. *Le rôle des organisations paysannes dans le système de recherche et de vulgarisation agricoles. Le cas des UCP de Kalalé et Boukoumbé et de l'ACooBéPA de Tchaourou/Ouèssè au Bénin.* Working Document KIT/FUPRO, Amsterdam/Bohicon, The Netherlands/Benin.

Notes

1. The ANFANI project (1998-2000), funded by The Netherlands Directorate General for International Cooperation (DGIS) within the framework of the Agreement for Sustainable Development.
2. The *Projet d'Amélioration et de Diversification des Systèmes d'Exploitation* (PADSE, 2000-2004), funded by the French Development Agency (AFD); see also AFD (2005: pp. 68-85).
3. INRAB has a regional research centre in Central Benin that manages the forestry research programme.
4. See the case study on UCPC Benin elsewhere in this publication.
5. The enterprise Anfani Garbi SARL was involved in the ANFANI project. Opinions differ about the actual reasons why this project didn't succeed.
6. The social share *(part sociale)* is a pre-defined amount of money that every member pays when becoming a member. It allows for constituting a fund (i.e., working capital) for the farmer group. This fund can be used to procure equipment or construct infrastructure, which then becomes collective, group property.
7. 1 USD = 545 francs CFA
8. 50% for the Association and 25% for the ULGPA and GPA each.
9. The *Nagot* are related to the *Yoruba* through migration from Nigeria.
10. A joking relationship is an interaction between two social groups whereby one group is allowed to make fun of the other, and where the latter is not permitted to take offence. It is an institutionalized mechanism that allows for criticizing each other and resolving conflicts.
11. The principal author of this case study didn't meet any female members who borrow or rent cashew plantations.
12. A price to be paid at the farm gate is defined before marketing the product. However, when actual selling on the world market takes places, prices may turn out to be higher. This allows for paying a separate, supplementary amount – the rebate – to producers.
13. For example the ICCO-funded *Programme de Développement Intégré* 2004-2005. See also: www.icco.eu/delivery/projectenboek/2004/continenten_projecten/BJ001101.html
14. *Groupements Villageois*; see case study on UCPC Benin elsewhere in this publication.

About the authors

Willem Heemskerk is a research and development management and organization specialist with 25 years of experience. His specific experience is in the field of facilitating demand-driven and client-oriented research and development with a systems perspective, both at national and sub-national level. He has worked for organizations such as DGIS, FAO and KIT, for national research and development organizations in Sao Tome and Principe, Mozambique, Zambia and Tanzania. His short-term assignments have been to Anglophone and Portuguese speaking countries in Sub-Saharan Africa, as well as to Brazil, Bolivia, Philippines and India. He is currently based in Amsterdam at the Royal Tropical Institute (KIT), Department of Development Policy and Practice (DEV), and undertakes short-term assignments in the fields of client empowerment and rural service provision.
E-mail: w.heemskerk@kit.nl
Website: www.kit.nl

Adolph Aloyce Kumburu is a registered professional process engineer, with additional training in management, agri-marketing and advanced business planning. He has a background of thirteen years experience in coffee industry as director of operations at Mbinga Coffee Mill, Tanzania, involved in milling, buying and selling coffee in auctions, and installation and operating coffee wet processing stations. He is also a coffee grower with approximately 6 hectares of coffee under cultivation and has served on numerous committees of the coffee industry in Tanzania. He is the Executive Director of KILICAFE since 2002. He is the spokesman of the farmers' association. Since 2002, he was a coffee business advisor of Technoserve Tanzania, managing the coffee programme in Ruvuma, Mbeya and Iringa Regions in Southern Tanzania. Currently he is the Coffee Programme Manager of Technoserve Tanzania, managing their national coffee program.
E-mail: adolph.kumburu@tnstanzania.org or edirector@kilicafe.com
Website: www.kilicafe.com

Richard Masandika is the network coordinator of *Mtandao wa Vikundi vya Wakulima wa Wilaya ya Monduli*, MVIWAMO (the Network of Farmer Groups in Monduli District), a middle network of MVIWATA. In this position, he is responsible for the overall coordination of support to member farmer groups of MVIWAMO. He has extensive experience with capacity reinforcement of farmer groups and institutional development for demand-driven service provision.
E-mail: mviwata-monduli@habari.co.tz

Suzanne Nederlof is rural development sociologist with experience in platform building, experiential learning, innovation systems, farmer-based organizations and agricultural research and extension. She has more than 9 years of working experience in Sub Saharan Africa where she worked in Burkina Faso for an outreach station of the Wageningen University in the Netherlands, the International Center for Soil Fertility and Agricultural Development (IFDC) in Togo, and the United Nations Food and Agricultural Organisation (FAO) in Ghana. She holds a PhD in Communication and Innovation Studies. During the period of her PhD she worked with the Convergence of Sciences Programme and did research on how agricultural research could become more beneficial to resource-poor farmers in Sub Saharan Africa. She is presently employed as an advisor in sustainable economic development at the KIT Department of Development Policy and Practice in Amsterdam.
E-mail: s.nederlof@kit.nl
Website: www.kit.nl

Jean Damascène Nyamwasa is an agronomist by training who started his professional career in 1988 with management positions in services for rural development in Rwanda. From 1991 till 1996 he occupied several positions in the training and advisory department of the Centre IWACU, which is a Centre for training and co-operative services in both the agricultural and non-agricultural sector. From 1997 untill now he operates as a free lance consultant in rural development and co-operative management. He did numerous consultancy, training and facilitation assignments for donors, national and international NGOs and development co-operation agencies with a focus on farmers' organizations and rural services for farmers. Since 2000 he is also a technical advisor for the Dutch Agriterra foundation and as such responsible for technical back-up of their support programme of farmers' organizations in Rwanda.
E-mail: planeef@rwanda1.com or nyamwasa2@yahoo.fr

Stephen Ruvuga is a rural development specialist who trained in agronomy and ecological agriculture. He has sixteen years experience in rural development. He worked as development specialist at the SADC-ICRAF Agroforestry Research and Development Project in Tanzania, as a senior trainer at the INADES Formation Tanzania, and as agricultural reseach officer at the Farming Systems Research

Programme at Ilonga Agricultural Reseach Institute. Currently he is the National Coordinator of *Mtandao wa Vikundi vya Wakulima Tanzania*, MVIWATA (the National Network of Farmers Groups in Tanzania).
Email: mviwata@africaonline.co.tz

Clarisse Tama-Imorou is a rural sociologist and anthropologist from Benin and member of the European and African association for anthropology of social change and development (APAD). Since the start of her professional career in 2000, she worked as a free-lance consultant for a community-driven development project and a support project for farmers and women farmers in Northern Benin. She is also an experienced trainer and facilitator who worked with women processors groups and on gender issues. In addition, she continued to conduct action-research on farmers' strategies and impact of rural development projects. In 2004 she was a laureate for the small grants programme of the Council for the Development for Social Science Research in Africa (CODESRIA) for her work on social differentiation in Northern Benin. She is currently involved in PhD research on primary education through the German Academic Exchange Service (DAAD).
E-mail: clartama@yahoo.fr

Bertus Wennink is a specialist in multi-stakeholder approaches for institutional development, management of demand-driven agricultural services, capacity strengthening of farmer organizations, community driven development, and natural resource management. He is an experienced advisor on institutional and organizational development for enhancing demand-driven services for pro-poor development. He joined KIT in 1994 and now has 18 years of experience, mainly through long-term assignments for SNV and KIT in Mali and Benin and short-term expert missions in West and Central Africa for donors, government organizations, including National Agricultural Research Institutes and Farmer Unions. Since 2003, he is based at the KIT Department of Development Policy and Practice in Amsterdam as a senior researcher, advisor and trainer.
E-mail: b.wennink@kit.nl
Website: www.kit.nl

Colophon

Bulletins of the Royal Tropical Institute (KIT)
The KIT Bulletin Series deals with current themes in international development. It is a multi-disciplinary forum for scientists, policy makers, managers and development advisors in agriculture, natural resource managements, health, culture, history and anthropology to present their work. These fields reflect the broad scope of KIT's activities.

KIT Development Policy & Practice
KIT Development Policy & Practice is the Royal Tropical Institute's main department for international development. Our aim is to contribute to reducing poverty and inequality in the world and to support sustainable development. We carry out research and provide advisory services and training in order to build and share knowledge on a wide range of development issues. We work in partnership with higher education, knowledge and research institutes, non-governmental and civil society organizations, and responsible private enterprises in countries around the world.

Contact information
Royal Tropical Institute (KIT)
KIT Development Policy & Practice
KIT Publishers
PO Box 95001
1090 HA Amsterdam
The Netherlands
Telephone: +31 (0)20 568 8458
Fax: +31 (0)20 568 8444
Email: b.wennink@kit.nl
Website: www.kit.nl/development

© 2007 KIT, Amsterdam, The Netherlands

This is an open-access publication distributed under the terms of the Creative Commons Attribution License, which permits unrestricted use, distribution, and reproduction in any medium, provided the original author and source are credited.

Edited by Barbara Shapland
Cover and design Studio Berry Slok, Amsterdam, The Netherlands
Cover photo Roel Burgler
Printing High Trade NV, Zwolle, The Netherlands

Correct citation
Wennink, B., S. Nederlof and W. Heemskerk (eds.) (2007) *Access of the poor to agricultural services: the role of farmers' organizations in social inclusion.* Bulletin 376. Amsterdam: KIT Publishers.

Keywords
Agricultural development, Farmers' organizations, Innovation, Rural innovation, Rural livelihoods, Service delivery, Sub-Saharan Africa

ISBN: 978 906832 642 0
ISSN: 0922-7911
NUR: 600/940

Partner organizations for this publication

Fédération des Unions de Producteurs du Bénin (FUPRO)
Website: www.fuproben.org

Assocation of Kilimanjaro Specialty Coffee Growers (KILICAFE), Tanzania
Website: www.kilicafe.com

Mtandao wa Vikundi vya Wakulima Tanzania (MVIWATA)
Email: mviwata@africaonline.co.tz

Mtandao wa Vikundi vya Wakulima wa Wilaya ya Monduli (MVIWAMO)
Email: mviwata-monduli@habari.co.tz

Syndicat Rwandais des Agriculteurs et Eleveurs (INGABO)
Email: ingabo@rwanda.1.com

Other publications in the Agricultural Innovation Systems Series

- Heemskerk, W., N. Lema, D. Guindo, C. Schouten, Z. Semgalawe, H. Verkuijl, B. de Steenhuijsen Piters and P. Penninkhoff. 2003. *Un guide pour la recherche agricole régie par la demande.* Approche Gestion de Recherche Orientée Client (AGROC). IER, Mali; DRT, Tanzanie; et KIT, Amsterdam. KIT-Publishers, Amsterdam.
- Heemskerk, W., N. Lema, D. Guindo, C. Schouten, Z. Semgalawe, H. Verkuijl, B. de Steenhuijsen Piters and P. Penninkhoff. 2003. *A guide to demand-driven agricultural research.* The Client-Oriented Research Management Approach (CORMA). IER, Mali; DRT, Tanzania; et KIT, Amsterdam. KIT-Publishers, Amsterdam.
- Heemskerk, W. en B. Wennink. 2004. *Building social capital for agricultural innovation.* Experiences with farmer groups in Sub-Saharan Africa. Bulletin 368. KIT-Publishers, Amsterdam.
- Heemskerk, W. and B. Wennink. 2005. *Stakeholder-driven funding mechanisms for agricultural innovation. Case studies from Sub-Saharan Africa.* Bulletin 373. KIT-Publishers, Amsterdam.
- Wennink, B. and W. Heemskerk (eds.), 2006 *Farmers' organizations and agricultural innovation. Case studies from Benin, Rwanda and Tanzania.* Bulletin 374. KIT-Publishers, Amsterdam.